NEW YORK REVIEW BOOKS
CLASSICS

ONWARD AND UPWARD IN THE GARDEN

KATHARINE S. WHITE (1892–1977) was born in Winchester, Massachusetts, the youngest of three daughters. She attended Miss Winsor's School and Bryn Mawr College, and in 1915 married Ernest Angell, with whom she had two children, Nancy and Roger. She became the first fiction editor at *The New Yorker* in 1925. Four years later, she met and, after separating from her first husband, married E. B. White, with whom she had one son, Joel. In the early 1930s the Whites bought a farmhouse in North Brooklin, Maine, and by the end of the decade they moved there from New York. White began writing garden pieces for *The New Yorker* in 1958, in the waning years of her long career as fiction editor, in which she exerted a profound influence on twentieth-century American literature. *Onward and Upward in the Garden* (1979) is her only book, edited and published posthumously by her husband E. B. White.

E. B. WHITE (1899–1985) was the youngest of six children, born in Mount Vernon, New York. He attended Cornell University, where he earned his lifelong nickname, Andy. He worked as a reporter and a copywriter before being recommended for a job at the newly founded *New Yorker* by the fiction editor Katharine Angell, who had read his submissions to the magazine. White became a regular contributor to *The New Yorker* in 1927, writing essays known for their humor and honesty. He was the author of three beloved children's books, *Stuart Little* (1945), *Charlotte's Web* (1952), and *The Trumpet of the Swan* (1970), and the co-author of *Elements of Style* (1959) with his former professor William Strunk Jr. In 1971 he received the National Medal for Literature.

ONWARD AND UPWARD IN THE GARDEN

KATHARINE S. WHITE

Edited and with an introduction by

E. B. WHITE

NEW YORK REVIEW BOOKS

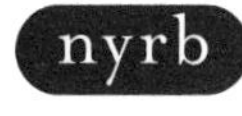

New York

THIS IS A NEW YORK REVIEW BOOK
PUBLISHED BY THE NEW YORK REVIEW OF BOOKS
435 Hudson Street, New York, NY 10014
www.nyrb.com

The text of this book originally appeared in *The New Yorker.*

Library of Congress Cataloging-in-Publication Data
White, Katharine Sergeant Angell, author.
Onward and upward in the garden / by Katharine White ; edited and with an introduction by E. B. White.
pages cm. — (New York Review Books classics)
Originally published: New York : Farrar, Straus, Giroux, copyright 1979.
ISBN 978-1-59017-850-8 (alk. paper)
1. Gardening—Book reviews. 2. Seeds—Catalogs—Book reviews. 3. Nursery stock—Catalogs—Book reviews. 4. Gardens—Book reviews. 5. Flower arrangement—Book reviews. I. White, E. B. (Elwyn Brooks), 1899–1985, editor. II. Title. III. Series: New York Review Books classics.
SB453.W4268 2015
635.02'16—dc23
2014038282

ISBN 978-1-59017-850-8
Available as an electronic book; 978-1-59017-851-5

Printed in the United States of America on acid-free paper.
10 9 8 7 6 5 4 3 2 1

Contents

Introduction

In its issue of March 1, 1958, under the heading BOOKS, *The New Yorker* ran a critical article on garden catalogues. It was signed Katharine S. White, its subhead was "Onward and Upward in the Garden," and it was the first of what was to become a series of fourteen garden pieces extending over a period of twelve years.

To readers of the magazine, this first piece came as a surprise—I think a pleasant one. It was innovative in that its author, without warning or apology, plunged boldly into reviewing the books of seedsmen and nurserymen as though she were reviewing the latest novel. To the best of my knowledge, nobody had ever done this before in a magazine of general circulation. If *The New Yorker*'s readers were surprised, imagine the condition of the seedsmen and nurserymen themselves! It must have been a dream come true for them to wake up and discover that their precious catalogues and their purple prose were being examined by a critical mind in the pages of a well-regarded publication.

Once having taken off, Katharine White lost no time in slipping into a critical mood. "Say you have a nice flower like the zinnia," she wrote, ". . . clean-cut, of interesting, positive form, with formal petals that are so neatly and cunningly put together, and with colors so subtle yet clear, that they have always been the delight of the still-life artist. Then look at the W. Atlee Burpee and the Joseph Harris Company catalogues and see what the seedsmen are doing to zinnias." There followed a gentle rebuke, directed at the trend toward making flowers bigger and fancier. From that moment, she was off and running. Seedsmen, hybridizers, nurserymen hung on to their hats and read about themselves—their triumphs, their failures, their conceits, their prose style—with a mixture of awe, thanksgiving, and skepticism. They were being scrutinized in public by a woman who, whatever else she was, was an opinionated gardener, a devoted lover of flowers and vegetables, and an addict of catalogues. She was also well along in years and had arrived at her opinions over a long period of time and after much gardening.

In addition to surprising thousands of *New Yorker* readers and dozens of seedsmen, Katharine managed to startle a third party—me, her husband. I felt both surprise and apprehension at this curious and unexpected development. I had observed my wife as a gardener but had never thought of her as an authority on horticultural matters, although she was clearly a devotee; I did not regard her as a scientist, and I wondered whether she realized what she was getting into. But I did know that she need bow to no one in her love of flower form and her devotion to the spirit and the task of gardening.

Writing, for her, was an agonizing ordeal. Writing is hard work for almost everyone: for Katharine it was particularly hard, because she was by temperament and by profession an editor, not a writer. (The exception was when she wrote letters. Her letters—to friends, relatives, contributors—flowed naturally from her in a clear and steady stream, a warm current of affection, concern, and eagerness to get through to the mind of the recipient. Letters were easy. How I envied her!) But when she sat down to compose a magazine piece on gardening, faced with all the strictures and disciplines of formal composition and suffering the uneasiness that goes with critical expression in the public print—this was something else again. Gone was the clear and steady stream. Katharine's act of composition often achieved the turbulence of a shoot-out. The editor in her fought the writer every inch of the way; the struggle was felt all through the house. She would write eight or ten words, then draw her gun and shoot them down. This made for slow and torturous going. It was simple warfare—the editor ready to nip the writer before she committed all the sins and errors the editor clearly foresaw. Occasionally, I ribbed her about the pain she inflicted on herself. "Just go ahead and write," I said. "Edit it afterwards—there's plenty of time." My advice never had any effect on her; she fought herself with vigor and conviction from the first sentence to the last, drawing blood the whole way. I worried about her health, which was not good. Luckily she always did her homework and never started to write anything until she knew what she was talking about.

In bringing these pieces together in a book, I draw encouragement from a number of sources. First, she herself had hoped to bring out a garden book. She kept putting the matter off because she wanted to expand her existing articles and write one more piece—a reminiscence about the gardens of her childhood. Second, a number of people whose judgment I valued had urged Katharine to put her material into a book. Marianne Moore was, I think, the first one to persuade her that the *New Yorker* pieces should be preserved between covers. Finally, when I got round to reading the articles, I discovered to my great joy that though they began twenty years ago they still seemed fresh. There had been, of course, many changes in the cast of characters and in the business firms, but there was a quality in her work that the years did not diminish: her excitement at growing things, her belief in the worth of gardening, her fascination with the architects who were behind the catalogues—these were timeless. She held the whole world of flowers in a warm embrace. Her pieces were intensely personal and charged with emotion.

To write of Katharine simply as a gardener would be like writing of Ben Franklin simply as a printer. Gardening was indeed a part of her, but it was never her major interest, consuming all her thoughts and all her talents. She simply accepted the act of gardening as the natural thing to be occupied with in one's spare time, no matter where one was or how deeply involved in other affairs. "Although I have gardened all my life," she wrote to a reader in 1959, "I am very much of an amateur and I am only beginning to know a little about the many nurseries that specialize."

Amateur she was, but although she lacked special knowledge and a background of science, she was bountifully endowed with strong feelings, a sure taste for what was simple and good, and the excitement and the fun that came with discovery. She had a fierce loyalty to common gifts of nature: goldenrod, pussy willows, dandelions, violets, wild flowers. As she began evaluating the catalogues of the professionals, she was enchanted to find that these articulate nurserymen were not just names on a page, they were live actors on her chosen stage; they were writers, and they were a special breed of cat—stylists of sorts, quarrelsome, opinionated, outspoken, and loaded with exact information and personal bias. They fascinated her but failed to intimidate her, and when Amos Pettingill, the sage of White Flower Farm, whose name sounded a bit too Dickensian for her taste, described in his catalogue a French pussy willow as "not the unreliable wild Pussy Willow," this was too much for my New England wife. "What is unreliable, pray, about the native wild pussies?" she retorted. "I have found them trustworthy in every respect."

I did not know at the time, and still don't know, what moved Katharine to write that first garden piece. Ever since I'd known her, she had been surrounded by seed catalogues in the wintertime, by seedlings pushing up in flats in spring, and she had always arranged bouquets for our house or apartment, whether the blooms came out of her own garden in Maine or out of a florist's shop in town. But no writing of all this had occurred. In 1958, her job as an editor was coming to a close and this provided her with more time to look about, more time to think about the gardens of her life. I suspect, though, that the thing that started her off was her discovery that the catalogue makers—the men and women of her dreams—were, in fact, writers. Expression was the need of their souls. To an editor of Katharine's stature, a writer is a special being, as fascinating as a bright beetle. Well, here in the garden catalogues, she stumbled on a whole new flock of creative people, handy substitutes for the O'Haras, the Nabokovs, the Staffords of her professional life. I imagine this was what did it—she couldn't stay out of the act any longer. She began reading Will Tillotson, Cecil Houdyshel, Amos Pettingill, Roy Hennessey ("Oregon's angry man"), H. M. Russell (the day-lily man of Texas), and many others. She was out of the *New Yorker* office but back among writers again, and in a field that had endless allure for her—the green world of growing things. "Reading this literature," she wrote, "is unlike any other reading experience. Too much goes on at once. I read for news, for driblets of knowledge, for aesthetic pleasure, and at the same time I am planning the future, and so I read in dream."

"Onward and Upward," a phrase swiped from the Unitarian creed, had often appeared in *The New Yorker* as a heading in other contexts. It was Katharine's invention, however, and it proved a happy choice of title for her adventure in garden writing. After two years of giving the catalogues an annual going-over and assessing the prose style and idiosyncrasies of the writers, she set to work, onward and upward, exploring other aspects of the garden world: the history of gardens, the literature of gardens, the arranging of flowers, the herbalists, the trends and developments. She was blessed with a curious mind and she ranged widely. She continuously revealed her own prejudices, her likes and dislikes, her crotchets. Katharine was a traditionalist, not only in the garden, but everywhere else. She preferred the simple to the ornate, the plain to the fancy, the relaxed to the formal, the single to the double, the medium-size to the giant. She detested abbreviations: snaps, mums, glads, dels. She did not care for flowers that, to her eye, were gross or pretentious or stiff. (She seldom grew dahlias or gladiolas on our place.) She loved old clay pots, despised new plastic ones—the "hideous green, lavender, or gray plastic pot, often striated with pink." She knew where she stood, and she was not a woman who looked with indulgence on a pot striated with pink.

Studying the Burpee catalogue (1959), she gazed fearfully ahead to the time when our garden beds would be full of great shaggy heads, "alike except for color, all just great blobs of bloom." When she discovered a zinnia named Miss Universe, seven inches across, she was reminded of her own plight as a girl, when a relative informed her that the great knot of hair on the back of her head was a deformity. She deeply loathed artificial flowers—considered them a crime against Nature; and once, reading in the papers that the city of Washington was to be favored with some artificial shrubbery, she dashed off a letter of complaint to Ladybird Johnson. The newspaper report turned out to be false, and the reply she received from the First Lady was a model of friendly reassurance and moral support.

Not only was she traditional by nature, Katharine showed a strong streak of parochialism in her approach to gardening. New England was what she knew as a child, and the roots of her ancestors went deep in the soil of Maine and of Massachusetts. The things that grew in New England, therefore, were "correct." They occupied a special place in her heart, an authenticity not enjoyed by flowers that made the mistake of blooming in other sections of the country. In my bumbling way, I ran straight into this regional bias when I introduced a few camellia plants in the small greenhouse we had acquired. This was in the last years of her life, when her days were becoming increasingly dreary because of invalidism and I was casting about for ways to give her a few scattered moments of pleasure against the sombre backdrop of illness. I must have seen, years before, what she had written about camellias: "My eye admires them even while my taste rejects them . . . I live in an old New England farmhouse, against which these lush blooms look out of period." I read it, I guess, but it must have gone out of my mind. So I sent off for several camellia plants, thinking to cheer her up. The plants arrived from the West Coast by Greyhound Bus and soon came into bloom—Tiffany, Betty Sheffield Supreme, Ville de Nantes, and other handsome little things. Proud of myself, I would make a quick, early trip to the greenhouse, come back with a single flower, put it in a tiny vase, and have it on the tray when I carried her breakfast up to her room—just a small matutinal surprise, courtesy of the management.

"Oh! How pretty!" she would exclaim. But Katharine was never any good at concealing her true feelings—she would have been the world's worst poker player. I soon detected, on these early-morning visits, a forced note in her voice. I could almost hear her muttering under her breath, "Now, what is a Southern flower doing in my Maine bedroom?"

How she loved shopping in catalogues! Hour after hour she studied, sifted, pondered, rejected, sorted—in the delirium of future blooming and fruiting. Harris was her dream catalogue, it was always within reach. No longer able to sit at a desk or at a typewriter, she had abandoned her cozy study at the front of the house and taken up a place at one end of the living-room sofa, propped with pillows. This became the control center of the house. The sofa served as desk as well as seat and it soon became buried under a mountain of catalogues, books, letters, files, memoranda, Kleenex, ash trays, and miscellany. The extraordinary accumulation, which would have driven me crazy, never seemed to annoy her or slow her up. I built her a coffee table, to catch the overflow from the sofa. The table was soon groaning under its own load. Yet she usually knew where something was, however deeply it was buried.

Although she spent a lot of time in dreamy admiration of the seductive pictures in catalogues—the Impossible Tomato, the Ultimate Rose—she was actually a hardheaded planner and organizer. She did little of the physical work of gardening herself; in youth she lacked the time, in age she lacked the strength. Henry Allen, our caretaker and himself an ardent gardener, was her strong right arm. But she masterminded everything. She got her seed orders in early, sometimes directing the campaign from her bed in a hospital, and we always ended up with a vegetable garden that loaded our freezer and nourished our bodies, and with flower borders that filled the eye and the spirit. She shopped among the seedsmen lavishly but cannily.

She also added them to her already long list of personal friends. Soon she knew intimately a great many of the people behind the catalogues, and when she sent in a seed order or a plant order it was usually accompanied by a long chatty letter in which she gave a quick rundown of the doings on our own place and then sought news of the trials, illnesses, and problems of the recipient. The garden people were quick to respond. "Dear Mrs. White," one letter began. "I am very happy to hear from you again, but so sorry you now feel old and ill. . . . I fractured my shoulder a couple of years ago, and somewhere along the way I lost my sense of equilibrium." And so the letters went, back and forth. I got the impression that my wife was in close touch with about half of the professional gardeners in America and worried about all of them.

Katharine never belonged to a garden club. I don't think she would have fitted in very well. In fact, had she joined, there is a good chance she would have been expelled for insubordination: she refused to pay any attention to the National Council with its dicta governing the acceptable arrangement of flowers in a container. Her garden was her clubhouse, where there were bugs but no rules.

When she got round to writing about the flower arrangers, she joyfully and conscientiously read all the books she could get her hands on. (Zen and all zat.) She was delighted to discover that flowers were being arranged fourteen centuries before Christ and that the Ikenobo School in Japan (founded in 1462) has three million students. She examined the "Principle of Three"—*shin* for Heaven, *so* for Man, *gyo* for Earth. Symbolism, however, was foreign to Katharine's nature; she never introduced heaven, man, or earth into our home—just petunias, salpiglossis, and snapdragons. She had long since listened to Lafcadio Hearn on the vulgarity of Occidental flower arranging but was not impressed. She felt that the *ikebana* vogue was impractical or unwholesome in the West: Americans were too impatient to create Japanese arrangements. She held steadfastly to her childlike ideal of bringing flowers indoors and getting them into place without a lot of fuss. She was not without ritual, but it belonged to no known religion. In her own way, Katharine was as bemused in the presence of flowers as any Oriental. She rejected symbolism but revered the beauty of flower form and the spiritual impact of the natural world as it was manifested in flowers.

The judging of "standard" shows by accredited judges who had undergone the rigid training prescribed by the National Council of State Garden Clubs got her back up. She had no use for any Word coming down from on high. Once, she happened to be present at a show and discovered a judge giving advice to a novice: "You really should *not* use Baby's Breath. It would never be accepted in a standard show." This was the Word. Katharine held no brief for Baby's Breath, but neither did she intend to sit still and allow *any* flower, whether it was Baby's Breath or ragweed, to be banished by national decree. She threw the Word right back.

One of the minor amusements I enjoyed for some forty years was watching my wife fix flowers, an operation that, in our home in Maine, usually took place a few minutes before lunch. Newspapers would be spread out on a round yellow table on the north porch—our summer dining room. Vases would be set out, and Katharine would arrive, fresh from the cutting garden, bearing baskets of assorted blooms, maybe seventy-five flowers in all, from which four or five arrangements would be built, to fit the *tokonomas* of our house. She had picked the flowers not at random but with an uncanny accuracy—the right kinds, the right length of stems, the right colors—to fulfill exactly some deep mental image she had of the final results. She never hesitated, she never fussed, and she was quite rough with flowers, as if to say, "If you can't take the heat, go away somewhere and wilt." She worked quickly, deftly, and seemingly without plan. This last was not true: her head was full of blueprints, images of the finished bouquets, and she worked in bold strokes to achieve them—a style I'm sure would have amazed the judges of the National Council. Sometimes, as I sat quietly in my corner, watching her throw flowers at each other, it looked as though she were playing darts in an English pub. Whenever I could, I attended these flower-arranging sessions (they lasted only ten or fifteen minutes), because it was a little like going to a magic show. She seldom spoke during the show, seldom commented on the finished product. Once in a while she would make a pronouncement, *sotto voce*, "There! That's pretty." But there was always the hint of a question mark buried in there. "*Is* it? Is it pretty?"

For Katharine, a room without a flowering plant was an empty shell. After she became so weakened by failing health that she required constant nursing, I equipped her bedroom with a hospital bed, a hospital table that spanned the bed, an oxygen kit, and a wonderful green device called a Bird Respirator. The patient gazed quizzically at this orderly and aseptic scene and immediately countered by calling for a Bird of Paradise plant in a tub. It was duly brought in and was about the size of a Shetland pony. Soon it was producing its bizarre parrot-like blossoms. It was a lecherous old thing, always grabbing at the nurses as they went by. We finally arranged matters so the Bird Respirator, when not in use, could sit lurking behind the Bird of Paradise, out of sight. It all worked out very nicely. The room was acceptable again.

When Miss Gertrude Jekyll, the famous English woman who opened up a whole new vista of gardening for Victorian England, prepared herself to work in her gardens, she pulled on a pair of Army boots and tied on an apron fitted with great pockets for her tools. Unlike Miss Jekyll, my wife had no garden clothes and never dressed for gardening. When she paid a call on her perennial borders or her cutting bed or her rose garden, she was not dressed for the part—she was simply a spur-of-the-moment escapee from the house and, in the early years, from the job of editing manuscripts. Her Army boots were likely to be Ferragamo shoes, and she wore no apron. I seldom saw her *prepare* for gardening, she merely wandered out into the cold and the wet, into the sun and the warmth, wearing whatever she had put on that morning. Once she was drawn into the fray, once involved in transplanting or weeding or thinning or pulling deadheads, she forgot all else; her clothes had to take things as they came. I, who was the animal husbandryman on the place, in blue jeans and an old shirt, used to marvel at how unhesitatingly she would kneel in the dirt and begin grubbing about, garbed in a spotless cotton dress or a handsome tweed skirt and jacket. She simply refused to dress *down* to a garden: she moved in elegantly and walked among her flowers as she walked among her friends—nicely dressed, perfectly poised. If when she arrived back indoors the Ferragamos were encased in muck, she kicked them off. If the tweed suit was a mess, she sent it to the cleaner's.

The only moment in the year when she actually got herself up for gardening was on the day in fall that she had selected, in advance, for the laying out of the spring bulb garden—a crucial operation, carefully charted and full of witchcraft. The morning often turned out to be raw and overcast, with a searching wind off the water—an easterly that finds its way quickly to your bones. The bad weather did not deter Katharine: the hour had struck, the strategy of spring must be worked out according to plan. This particular bulb garden, with its many varieties of tulips, daffodils, narcissi, hyacinths, and other spring blooms, was a sort of double-duty affair. It must provide a bright mass of color in May, and it must also serve as a source of supply—flowers could be stolen from it for the building of experimental centerpieces.

Armed with a diagram and a clipboard, Katharine would get into a shabby old Brooks raincoat much too long for her, put on a little round wool hat, pull on a pair of overshoes, and proceed to the director's chair—a folding canvas thing—that had been placed for her at the edge of the plot. There she would sit, hour after hour, in the wind and the weather, while Henry Allen produced dozens of brown paper packages of new bulbs and a basketful of old ones, ready for the intricate interment. As the years went by and age overtook her, there was something comical yet touching in her bedraggled appearance on this awesome occasion—the small, hunched-over figure, her studied absorption in the implausible notion that there would be yet another spring, oblivious to the ending of her own days, which she knew perfectly well was near at hand, sitting there with her detailed chart under those dark skies in the dying October, calmly plotting the resurrection.

—E. B. White

Acknowledgments

For their help in preparing this text for publication, I am indebted to the friendly librarians of the Horticultural Society of New York, the Massachusetts Horticultural Society, and the Bryn Mawr College Library. I am equally beholden to the New York Botanical Garden, for help with the illustrations. I am also grateful to four friends: Harriet Walden, Katherine R. Hall, Lorraine Hanson, and Roy Barrette, who assisted me in the work of editing, researching, and typing.

E.B.W.

To the Reader

Some of the nurserymen and seedsmen mentioned in these pages have since changed their place of business. Others have altered their way of doing business or changed their product. A few have disappeared from the scene.

For the reader's convenience, two lists have been provided at the end of the book. One shows the present status of firms mentioned in the text; the other indicates which books are still in print.

I

A Romp in the Catalogues

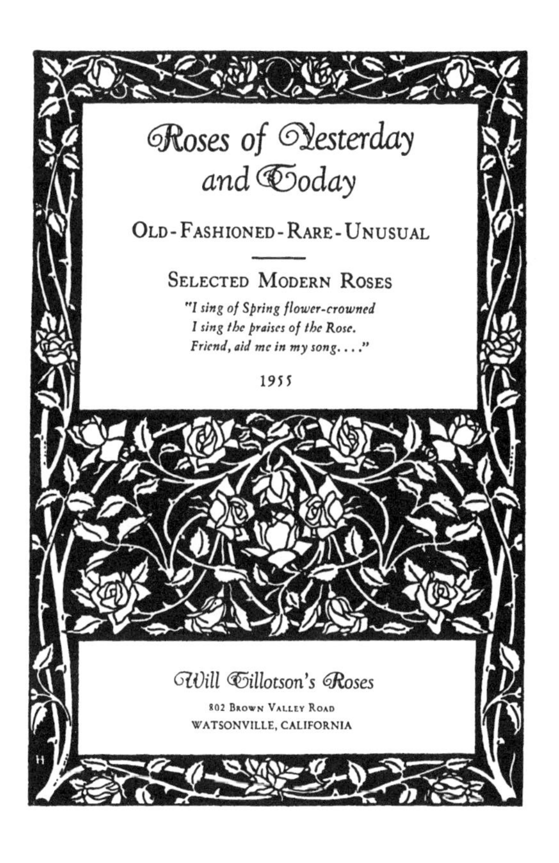
Roses of Yesterday and Today

OLD-FASHIONED-RARE-UNUSUAL

SELECTED MODERN ROSES

"I sing of Spring flower-crowned
I sing the praises of the Rose.
Friend, aid me in my song. . . ."

1955

Will Tillotson's Roses

802 BROWN VALLEY ROAD
WATSONVILLE, CALIFORNIA

March 1, 1958

For gardeners, this is the season of lists and callow hopefulness; hundreds of thousands of bewitched readers are poring over their catalogues, making lists for their seed and plant orders, and dreaming their dreams. It is the season, too, when the amateur gardener like myself marvels or grumbles at the achievements of the hybridizers and frets over the idiosyncrasies of the editors and writers who get up the catalogues. They are as individualistic—these editors and writers—as any Faulkner or Hemingway, and they can be just as frustrating or rewarding. They have an audience equal to the most popular novelist's, and a handful of them are stylists of some note. Even the catalogues with which no one man can be associated seem to have personalities of their own.

Before we examine the writers and editors, let us consider the hybridizers, and the horticulturists in general. Their slogan is not only "Bigger and Better" but "Change"—change for the sake of change, it seems. Say you have a nice flower like the zinnia—clean-cut, of interesting, positive form, with formal petals that are so neatly and cunningly put together, and with colors so subtle yet clear, that they have always been the delight of the still-life artist. Then look at the W. Atlee Burpee and the Joseph Harris Company catalogues and see what the seedsmen are doing to zinnias. Burpee, this year, devotes its inside front cover to full-color pictures of its Giant Hybrid Zinnias, which look exactly like great, shaggy chrysanthemums. Now, I *like* chrysanthemums, but why should zinnias be made to look like them? From Harris, you can buy the seed of what it calls New Super *Cactus* Flowered Zinnias, and they certainly do look like cactuses, or you can buy the seed of Fantasy Zinnias, which are the counterpart of asters. And both companies offer zinnias that look like dahlias. It is all very confusing. The Burpee people, who have always been slightly mad on the subject of marigolds, this year devote their outside front cover to their New Giant Fluffy Marigolds, which they describe as "large, round, fluffy, double chrysanthemum-like blooms"; this is just what they appear to be—chrysanthemums, not marigolds. In the Harris book, similar marigolds go by the *name* of Mum Marigolds. Meanwhile, what is happening to chrysanthemums? Well, some of *them* are being turned into cactuses, too; Wayside Gardens offers plants of one it describes as "Bronze Cactus—rare new mums with new style flowers." What is happening to cactuses this year I can't say, since I don't belong to the Succulent Society (a very active group, by the way) and have no cactus catalogues at hand, but I do own—among my other cactuses—a cactus a friend gave me years ago called a *poinsettia* cactus, and a miserable old thing it is, full of mealy bugs.

It is not only these transmogrifications of old favorites that trouble me; I am equally bothered by the onward-and-upward cry of the seedsmen and plant growers. Of course, this is nothing new; the trend since almost the beginning of time has been to grow things that are bigger and better. Better I go for: roses that blossom all summer, day lilies that stay open longer, lettuce with less tendency to bolt, corn that will not wilt, string beans without strings. The hybridizers—with an assist, perhaps, from the chemists—are responsible for all these blessings. But as for flowers, I have never been able to persuade myself that the biggest blooms are necessarily the most beautiful. Take the rose called Peace—"the rose of the century," one cataloguer terms it. Everybody knows this huge, rosy-yellow rose, and nearly everybody admires it and tries to grow it. In spite of its lovely colors, I don't like Peace. Even a small vaseful of Peace roses is grotesque, and on the bush the blossoms look to me like the cabbagy Tenniel roses of the Queen's Croquet Ground—the white ones Alice found the card gardeners hurriedly painting red against the arrival of the Queen. Lewis Carroll was prophetic; today the garden men are quite as busy changing the colors of flowers as they are changing their size and shape. Some years back, the Burpee breeders, in their zeal to produce a marigold with no gold in it, went to the length of offering their customers an award of ten thousand dollars for a marigold seed that would produce a white flower. (I suppose they were hoping for a sport.) In the last couple of years, they have cooled off a bit. The offer still stands, but instead of leading the catalogue it is farther back in the book. In the meantime, the company has developed a seed called Burpee's Nearest-to-White Marigold. I grew some of it last year, hoping, like everyone else, for a particularly white flower that would win the big prize.* Mine all turned out to be a sickly cream, which did not go well with the tawny oranges and golds and citron yellows of the common marigolds I always grow for autumn color. Even when the nurserymen aren't tampering with a flower, they like to appear to be doing so. One of White Flower Farm's proudest offerings this year is a shrub it lists as "White-Forsythia." Actually, it isn't a forsythia at all, it is a rare Korean shrub (*Abeliophyllum distichum*), which merely looks like forsythia, but without its golden charm.

Whatever may be said about the seedsmen's and nurserymen's methods, their catalogue writers are my favorite authors and produce my favorite reading matter. Most of them write anonymously, but a few men who sign their names offer fine examples of regional literature. Take the breezy and highly personal product of H. M. Russell, of Russell Gardens,

* In 1975, Burpee announced the winner of the White Marigold competition. A prize of $10,000 went to Mrs. Alice Vonk, of Sully, Iowa. The 1978 Burpee catalogue announces the "First White Marigold," and contains a promise by Mr. David Burpee that "we shall soon have other Whites, bigger and better."

Spring, Texas, a nursery that grows only day lilies. (Mr. Russell always makes it one word—"daylilies.") His 1955 booklet, which recently fell into my hands, starts out with "A Personal Note" to his friends and customers, thanking them for their confidence and reporting on his health, which had been poor but was better. This is followed by a two-page argumentative—you might say quarrelsome—essay headed "Concerning 'Evergreen' and 'Dormant' Daylilies and Their 'Regional' Behavior. Ours Are Hardy All Over America—Here's Why." Next comes a page of photographs showing four lily-growing Russells, and the biggest picture is of Mrs. H. M. Russell. "Without her inspiration," writes Mr. R., "there just wouldn't have been my contribution to the Daylily world." Farther on, amid the color pictures of the lilies, we are favored with a photograph of still another Russell—Skeeter, H.M.'s seventh grandchild, sitting behind his first birthday cake. Russell Gardens claims to be America's largest growers of day lilies; there are thirty solid acres of them, a great part of which, Mr. Russell says, "we consider strictly trash." I like this outspoken Texan; he may talk both big and small, but I find that I believe what he says. I hope he is in good health and I advise any fancier to investigate his lilies.

A stylist of another sort is nearer home—the sage of White Flower Farm, in Litchfield, Connecticut, who signs his writings "Amos Pettingill." I have no idea whether Amos Pettingill is a real person—the name sounds like an ill-advised fabrication—but, real or false, someone is in there pitching who edits a lively catalogue and writes in a highly distinctive style.* White Flower is a nursery that specializes in rare plants, shrubs, trees, and bulbs; it also grows good-quality everyday plants. Probably it is best known for its Warmenhoven strain of amaryllis, its clivia, its Blackmore & Langdon

* She soon found out. "Amos Pettingill" was William Harris, husband of Jane Grant, who was the first wife of Harold Ross. A small world.

tuberous-rooted begonias, and its *fraises des bois.* Its 1958 "Plant Book"—fifty cents a copy, thirty-seven pages—is a model of clarity, good taste, and good order, and (an oddity in garden catalogues) it is entirely in black-and-white and has no pictures of flowers, except for some decorative pen-and-ink drawings by Nils Hogner. The book is really a dictionary of White Flower's goods; everything offered is listed alphabetically under its botanical name and cross-indexed by its common name. (Only the catalogue addict will understand what a comfort so simple a thing as alphabetical order can be.) Mr. Pettingill contributes a slightly testy introduction, in which he takes off against the term "green thumb" and says that "top grade has nothing to do with size" (two ideas I subscribe to), then gets down to business on prices and what's new. To enjoy Amos Pettingill at his most peppery, though, you must read *White Flower Farm Notes*, which are bulletins issued at least six times a year and sent to White Flower's customers. (Non-customers may subscribe.) Here Amos really lets himself go, is chatty, sassy, or lyrical, as his mood dictates, and here he airs his latest enthusiasms and his pet grouches. ("BULBS, BULBS, BULBS. You will now [this was a June issue] please take your Plant Book in hand, and do what you had absolutely no intention of doing—order your bulbs for September-October delivery" and, elsewhere, "Confound it, we do not think customers are *always* wrong.") Most lady gardeners I know "simply love" him. I admire him more for his explicit cultural directions, his odd bits of gardening lore, and his sensible descriptions of what he has to sell than for the eccentric personality he has created for himself. I have found that what he says will happen to a plant usually does happen to it. White Flower grows good plants, packages them well, and sends exact and detailed directions (Amos again?) with every purchase. One of the new things it offers this spring, besides that white forsythia, is a lilac (*Syringa microphylla superba*) that blooms *twice* a year, and again it is featuring "Terra Rossa" unglazed clay flowerpots, hand-turned in Italy. These range from the eight-inch-diameter size, at $5.95, to the forty-inch, at $280—a nice little crock for the terrace.

Another writer with a style of his own is Will Tillotson, the grower and catalogue writer for Will Tillotson's Roses, Watsonville, California. The two catalogues I have—1955 and 1956—were lent me by a friend who runs a beauty parlor, and I must not keep them long, because though she has never bought a Tillotson rose, she reads Tillotson every night before she goes to sleep. (Catalogue readers, quite as much as catalogue writers, are a group apart.) The Tillotson catalogues are titled "Roses of Yesterday and Today" for the good reason that Will Tillotson, though he does grow and sell any rare or modern roses he happens to admire, specializes in old-fashioned roses. The catalogues are period pieces, too. The 1956 cover reproduces Sargent's "The Lady with the Rose," and 1955's is Fragonard's "The Swing," from the Frick Collection. I don't wonder that my friend finds these fat little books good to look at and to read. The many uncolored photographs of roses are enchanting, and I never tire of Mr. Tillotson's prose. He is a quoter. Bits of wisdom or poetry scattered through the 1956 book range from Yuan Chung-lang, a Chinese sage of the sixteenth century, to Leigh Hunt and Browning. The favorite source of Tillotson wisdom, though, is Dean Hole's *Book About Roses*, published in England in 1869. He quotes him again and again. The catalogue's descriptions of roses are informative and occasionally rhapsodic; if the date of a rose's origination is known, it is given. This will afford you an idea of the Tillotson style and scope:

OLD BLUSH. China. (1796) 4-5 feet, spreading. Not only "The Last Rose of Summer" as immortalized by the poet Moore, but also the first and in between, for this China rose literally never stops. A semi-double "fluttering assemblage of pink petals" giving an impression of airiness and gaiety. Don't plant it next to Chrysler Imperial (for instance), for "never the twain should meet."

Style and all, the Tillotson catalogues bring me great happiness. Here are the moss roses, the musk roses, the damasks and sweetbriers, the old-fashioned hybrid teas and hardy perpetuals of my childhood and my father's childhood, alongside rare or exotic roses and those modern floribundas and hybrid teas to which Will Tillotson gives the nod. Now and then, he stops carrying a rose he still thinks well of to make room for a new find—in 1956 he threw out Lowell Thomas—and sometimes he carries a rose he has not made up his mind about. I can't wait to learn whether in this year's catalogue he is still offering the rose called President Eisenhower; he has been wavering about it since 1954, when, he tells us, he wrote, "I will admit the rose is red, fragrant, forty-petalled, and is in nationally light supply for 1954. Beyond this I now refuse to go," adding that he was not yet ready to declare himself about either the rose or the man. This, he says, aroused a storm of protest, so in the summer of 1955 he wrote, "The C.W. [meaning the catalogue writer, Will Tillotson] has made up his mind about Pres. Eisenhower, the Rose (and the President of the U.S.)." A cagey statement if ever I saw one. For all I know, he is still sticking to it.

Of course, in New York, when ordering roses, one is likely to think first of the Jackson & Perkins Co., who announce themselves as the world's largest growers of roses, or of Bobbink & Atkins, who have been known to say they grow the world's largest *selection* of roses. Both statements are doubtless true. I have had good luck with J. & P.'s rosebushes, but I am increasingly disaffected by their catalogues, which each year grow more flamboyant. This spring's is a riot of huge color illustrations of the roses, interlarded with exclamatory type printed in red, green, yellow, blue, or black ink. It is a mess. I shall always be grateful to Jackson & Perkins, though, for years ago introducing me (and millions of others) to the New Yorker rose, which, I'm happy to say, considering its name, is the red rose that does not change color with age. My favorite among J. & P.'s yellows is still Eclipse, which is now listed under "Common Roses." It was not always so. But Eclipse's elongated pale buds, against its dark, glossy-green foliage, are as lovely as ever, and in my garden Eclipse continues to blossom long after the first frosts.

The current Bobbink & Atkins catalogue (1957, since the 1958 book is not yet ready) contains eighteen color pages of modern roses and three pages of what it calls "historical roses"—again the old-fashioned mosses and damasks, the French and the cabbage roses—and these listings are but a fraction of the hundreds of varieties the firm grows. On the cover is shown its new hybrid tea, a lusty, very clear pink rose with a satiny sheen, named in memory of Mrs. Cole Porter. To me, though, the loveliest rose pictured is Mermaid, a five-petalled climber or rambling bush rose with large pure-white-to-lemon blooms. This catalogue is not as inspiring in typography or general appearance as some of the firm's earlier ones, but it bears careful study if the reader is looking for unusual items, and it will convince him of the high quality of the plants and shrubs this nursery grows, even if he is not already aware of it from visits to the New York Flower Show, where B. & A. has always carried off many of the prizes. The catalogue's list of perennials contains some interesting items for rock gardens, such as hardy cyclamen, wild bleeding heart, and gentiana.

One of the most beautiful catalogues in my collection, and one of the best-edited, is Wayside Gardens' two-hundred-and-thirty-two-page book of plants, shrubs, vines, roses, bulbs, and other nursery stock. It offers the most varied and lavish collection of growing things I know of, and serious gardeners —especially those who go in for perennials—can hardly afford to be without it. The book is printed on coated stock, and more than half its pages are in full color—color that is extraordinarily lively and true. The typography has elegance, and the arrangement of the book makes it easy for the reader to find his way around. Most dear to the heart of the gardener, though, are Wayside's unexaggerated descriptions, its directions for culture, and its exactitude in giving the height of each species—the biggest help of all in planting. The prices, as might be expected, are not low.

Here are a few of Wayside's offerings that took my eye. First, the three full-page color photographs of day lilies, which are captioned "The Finest of All Yellow Lemon Lilies!," "The Finest Apricot Hemerocallis," and "The Finest Red Hemerocallis." (Did I say "unexaggerated"? A simple superlative is nothing in a catalogue.) The red—*Hemerocallis helios*, named for the sun god—is a standout, for the rich red has no hint of the usual day-lily rust in it. But prepare yourself for disappointment: Wayside is sold out of Helios for this spring; order now for September delivery. I was fetched, too, by the color picture of Allwood Brothers' Laced Pinks—white or pale-rose petals with red-rose lacings. These pinks were brought to Scotland by the Huguenot weavers when they were driven out of the Low Countries; the refugees, it seems, settled near Paisley, and there took to hybridizing their homeland garden pinks. (Wayside tells you things like this.) I also liked another importation from Allwood Brothers, who are England's foremost growers of dianthus and pinks. This is the dwarf dianthus, *Allwoodii alpinus*, mixed white to pale-pink petals rayed with darker pink.

It would be impossible to cover here the infinite variety of Wayside's offerings, but you might want to consider the fringy *hardy* fuchsia, a rarity, or, if you like gigantism, the new deep-salmon phlox, Sir John Falstaff, whose florets are two inches in diameter. There are, of course, pages and pages of roses, and most of them are standard varieties. I like Hill Top, which cannot be found in all rose catalogues—loosely rolled buff petals shading to pink—and which has done well for me in a cold climate. There are also color pictures of two lovely single hybrid teas—pale-pink Dainty Bess, and White Wings, silver white with rosy amethyst stamens.

Now for a few specialties. Peony growers are among the most special of the lot. Cherry Hill Nurseries, West Newbury, Massachusetts, a good general nursery, is famous for its peonies; generations of Thurlows have been propagating them. The last catalogue (1957) lists well over a hundred varieties of the doubles, the singles, and the Japanese type, and more are available on demand. The prices are moderate —$1 to $15—and I can testify that a dozen peony clumps I got from Cherry Hill in the early thirties still bloom their pretty heads off each June. The 1957 list contains no tree peonies, which are among the luxuries of the garden, but perhaps the firm has them. White Flower Farm does. I recommend, though, for the peony gardener who is up to it, the leaflet issued by Silvia Saunders, of Clinton, New York. Miss Saunders is continuing the hybridizing of peonies started by her father, A. P. Saunders, who was a professor of chemistry at Hamilton College. His is a name to conjure with in the world of peonies, and his daughter seems to be carrying on in his tradition. The leaflet she issued is very technical. In her listing of herbaceous hybrids, she gives the names of the two parents in each simple cross, and she offers triple and quadruple crosses as well. She also sells seeds, peony-species plants (i.e., the original strains), and tree peonies. The hybrids include not only the crosses usually shown in flower shows but those made between species not hitherto used, which have been obtained from botanical gardens and collectors in Europe and Asia. There is a set of these peonies that Miss Saunders calls the Mystery Group and that she describes in her catalogue as "Ivories, pearled shades, suffused mauves. Single to double." I quote the description to show that she hybridizes for beauty rather than for eccentricity; her work is an art as well as a science. Prices run from $5 to $50, but anyone who has seen the Saunders peonies in bloom, as I was once lucky enough to, will know that they are rare indeed.

This is not the time of year to buy most flowering bulbs, but some of the names to remember when you put in your summer orders for tulips, narcissi, daffodils, and other spring bulbs are Max Schling Seedsmen; P. de Jager & Sons, whose head offices are in Holland; and John Scheepers, Inc., another outlet for Dutch bulbs. The de Jager catalogue, printed in Holland, is a lovely thing—chaste, compact, and full of photographs, many of them superb examples of color printing. For lily bulbs of all kinds, Romaine B. Ware, of Canby, Oregon, issues the most attractive catalogue I have seen, and Oregon's lilies are famous. White Flower and Wayside also carry a big selection of bulbs.

The nursery catalogues, both general and special, may be absorbing, but the real spring excitement—for me, anyway—comes with the seed catalogues. I can't attempt to cover them all, and I shall limit my field to the three houses I usually buy from: the W. Atlee Burpee Company, the Joseph Harris Company, and Max Schling. I used also to study the catalogue of Breck's Seeds, which is a famous name in New England, the firm having been in business since 1818, but I use it now only for garden accessories. For all I know, its seeds are as good as ever, but the catalogue is now so full of what is headlined, in a flurry of capital letters, as "TREASURE TROVE of HANDY HELPS for easy GARDENING, easy LIVING!" that I haven't the patience to sort out the seeds from the gadgets. However, if you want to buy an automatic bird feeder or a Tahitian garden torch ("Light up outdoor get-togethers with Tahitian enchantment!"), Breck's is your answer.

In spite of that gay marigold cover and a few other color pages, the Burpee catalogue is not pretty. It contains a hundred and twenty-four pages of gray type and cloudy photographs, dimly printed on uncoated stock. Nevertheless, it is fun, if taxing, to read, and the reader is always aware of David Burpee, the president of the company, as both writer, chemist, and hybridizer. He leads off this year with a philosophical letter—"Civilization began in a garden . . ." and so on. The catalogue's emphasis is on flower seeds, to which it devotes twice as many pages as to vegetables. Burpee as a flower-seed house has one sterling quality: it remains loyal to the old-fashioned flowers even while tampering with them, and it still offers many old-timers not to be found elsewhere. Mr. Burpee makes almost as big a thing of petunias as he does of marigolds. You can get bedders, balcony types, fringed petunias, ruffled petunias—Giant Ruffled, that is. While I usually prefer the simpler forms of petunia, I would not be without one of Burpee's Giant Ruffleds—Mauve Queen, a petunia I have never found in any other catalogue. Though the color scheme may not sound alluring—light mauve or heliotrope on a white ground with purple veinings—Mauve Queen is a beauty.

The Burpee people go for ruffles in anything. To me a a ruffled petunia is occasionally a delight but a ruffled snapdragon is an abomination. The snapdragon is a very complicated flower form to start with, and it has style. Fuss it up and it becomes overdressed. Here is what Burpee has done to snapdragons:

Giant Ruffled Tetraploid Snapdragons . . . Tetra snaps are a Burpee scientific creation, originated through the treatment of the best regular or diploid varieties with colchicine—a drug or chemical derived from the bulbs of fall crocus. Genetically . . . they have twice the number of chromosomes in their cells as the regular . . . snapdragons.

The result is that the upper lip of each dragon's mouth is crinkled. (Hardly kissable.) All through the catalogue the word "tetra" appears, and all through the flowers the chromosomes have been multiplying, with varying results—greater size, more varied colors, fancier petals—and the end is not in sight. The happiest results seem to me to be among the sweet peas, of which Burpee offers a myriad—the Most Fragrant, the Early Flowering, the Giant Ruffled, the Heat-Resistant, etc. If you mix two or three of them, you can have sweet peas all summer.

And now to Harris, my dream catalogue. Vegetables come first here, and rightly, considering the reliability and extra something all Harris vegetables seem to possess. I could enlarge on the virtues of the corn named Wonderful, the Snowball Cauliflower, the Cornell Self-Blanching Celery, and the exquisite qualities and the differences of Salad Bowl, Oak Leaf, Bibb, and White Boston lettuce. Instead, I shall only praise Harris for the style, clarity, and good looks of its eighty-four-page catalogue. Printed on coated stock, larded with exceptionally good photographs, and arranged for ease of selection, it is a model of what a seed catalogue should be. Harris has greatly expanded its flower-seed offerings in recent years, and now also has a short list of plants and bulbs. All these are as handsomely presented as the vegetables are, except for four color pages of flowers, whose photography does not have the art of the black-and-whites; here, too, the editors seem to be following the cataloguers' way of all flesh by introducing red, blue, and yellow lettering. The Harris book is strictly anonymous, and no literary character impinges upon you with his folksy philosophy or his ideas on propagation, but whoever writes it does things I like; he may list a collection of frostproof flowers or a group suitable to grow for dried-flower arrangements, or he may issue a command: "Scatter seed of Iceland Poppies in your perennial garden this June." The seed packets, too, carry foolproof directions for culture. I have learned a lot about gardening from the Joseph Harris Company.

Max Schling's spring catalogue is short—thirty-nine pages—and its lists of seeds and plants are highly selective. No scare headlines, no large claims; moderation is the word. Schling's dahlias are merely "Meritorious." Hidden away in the small type of this book you can find unusual things: water lilies—day bloomers, night bloomers, and hardy; a *yellow* fuchsia called Swanley Yellow; the sunflower called Redwood Empire; camellias that will thrive in the North; seeds of the Percival delphinium, a white delphinium with a black eye. There is also a list of seeds for greenhouse culture.

If it is native wild-flower plants or ferns that you wish, you could send for the leaflet catalogue of Wake Robin Farm. It might be a rewarding experience to transplant into your garden or your woods some of its trailing-arbutus, lady's-slipper, or cardinal-flower plants. The Pearce Seed Company, a nursery for rare plants or the seeds of rare plants, offers native wild-flower seed, but only in packages of mixed varieties, divided into wild flowers for shade and wild flowers for sun. However, among its listings of rare seeds may be found some of the common wild flowers, like the bottle gentian and the trilliums; an exotic form of jack-in-the-pulpit hides there under the name of Himalayan Arisaema. The Pearce catalogue is no longer the collector's item it used to be, but it still makes good reading because it offers so many real rarities both for the garden and for growing in the house. These last range from the difficult orchids to such easy things to grow as the more unusual forms of geranium—Apple Blossom, Arcturus, and Bird's Egg.

P.S. This report must have a melancholy addendum. I've just received the new Tillotson rose catalogue (1957–58, two years in one). It is written and signed by Will Tillotson. But on the final pages of this second edition of the book there is a note to say that Mr. Tillotson died in England last summer, while he was there collecting roses for his California nursery. The business will be carried on by Mrs. Dorothy Stemler, who has long been active in it. In Will Tillotson's last list of stock, the rose President Eisenhower has been dropped, without comment.

"Oh, gentle reader, forgive me if you are 'sleepy' or bored or annoyed. The tired catalog-writer has emptied his 'sack of adjectives at your feet.' " These words, reprinted from 1955's "Roses of Yesterday," are used as Will Tillotson's sign-off in the new book. He is a writer gardeners will miss.

2

Floricordially Yours

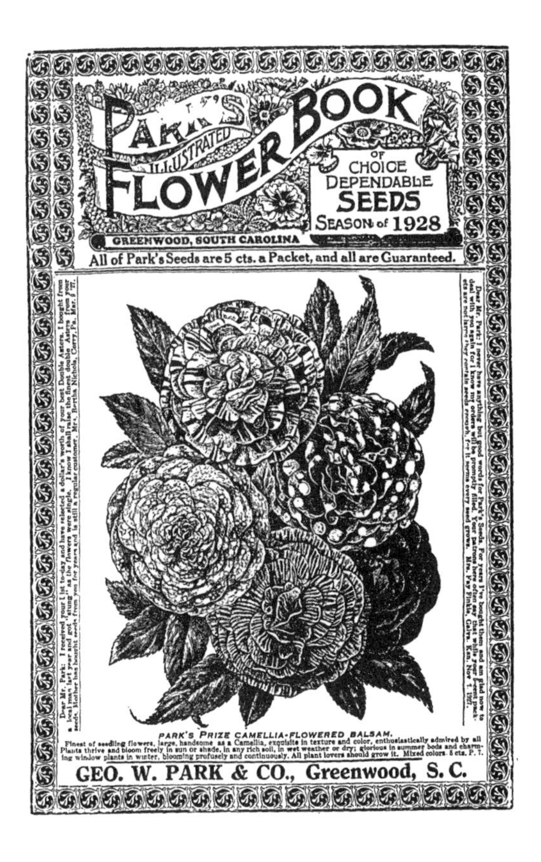

March 14, 1959

Amos Pettingill's introductory notes to White Flower Farm's annual "Plant Book" lead off this year with an emphatic plea to the customers to send their orders in early. First he cajoles, then he uses bait. Orders over ten dollars postmarked before February 15th win prizes and bonuses; orders postmarked between February 15th and March 25th get prizes but no bonuses. The scribe of Litchfield has not lost his disciplinarian tone. "It helps like anything," he tells his customers, "when we can read letters, without having to decode them. This is no Dead Sea Scrolls Operation."

The early order is not as easy for a gardener as Mr. Pettingill assumes. As I write, snow is falling outside my Maine window, and indoors all around me half a hundred garden catalogues are in bloom. I am an addict of this form of literature and a student of the strange personalities of the authors who lead me on. Gentle and friendly, eccentric or wildly vivid, occasionally contentious and even angry, every one of them can persuade me, because he knows what he is saying and says it with enthusiasm. Reading this literature is unlike any other reading experience. Too much goes on at once. I read for news, for driblets of knowledge, for aesthetic pleasure, and at the same time I am planning the future, and so I read in dream. Yet the present is naggingly with me, for I am in a state of torturing indecision. Will I, for example, have space in a south window next winter for a pot of Mr. J. N. Giridlian's "novelty of the year," *Habenaria radiata*, the Egret Flower, which this grower of rare bulbs is presenting for the first time in this country? Or should I make my new venture of growing an exotic plant one of Mr. Cecil Houdyshel's group of *Haemanthus*, the blood lilies of South Africa, "so rare that few have seen them"? Whichever I choose, it should be started now. Yet what I probably ought to decide is whether the household's major crop of beans for next summer's eating and freezing will be Harris's New Tendercrop, "the best flavored green beans," or whether I should, at less cost, remain loyal to the firm's Improved Tendergreen Beans, which my family and I have eaten with such relish through the past summer, fall, and winter. After all, Harris still praises *them*. And there is Pearl-green, too, with its pretty white seeds. But the beans can wait; they can't possibly go into our cold Maine ground until May. What I *really* must figure out at once, before others snatch up all the choice roses, is which varieties to reserve as replacements for the hybrid teas that have been killed by this severe winter. Yet how can I possibly know now how many will survive? Then comes the question of where, among the scores of excellent nurseries, to place my rose order. The rose growers, a particularly clamorous and competitive lot, are soon buzzing around inside my head like pestering gnats. At this rate, White Flower Farm will be lucky if it gets my order by May.

The catalogues induce a state of suspended animation; nevertheless, I shall try to report a few of the pleasures, discomforts, and discoveries of my 1959 garden reading. My happiest discovery, perhaps, is "Park's Flower Book," the catalogue of the Geo. W. Park Seed Company, of Greenwood, South Carolina. The charms of this friendly, crowded seed book will be no news to many catalogue readers, for I gather that most of the South, and a good section of the North as well, has been buying seeds from Park for the better part of a century. The firm sells a few vegetable seeds, too, but flowers come first.

The first impression the big volume gives is of soft, harmonious color. Though this is not a de-luxe publication, like that of Wayside Gardens, the color printing is exceptionally good and true to life, and the juxtapositions of the colors are pleasing. The opening page, headed "There Is Peace in the Garden," is full of many things. There are photographs of the founder (Geo. W. Park, 1852–1935) and of his sons, William J. and George B. Park, the latter now president of the company. There is a poem by Emerson titled "Musketaquid"—"All my hurts/My garden spade can heal"—and a prayer by John Henry Jowett of thanks for all beautiful things, including flowers, followed by the question "My friends, are these your sentiments? If not, why not?—PARK." There are two little Victorian engravings that must date from the very early catalogues. One is of a boy and girl and a wheelbarrow full of flowers. He pushes the barrow while she shields the flowers from the sun with her parasol. The caption reads "Yes, Sister, I buy only of Park; his seeds always grow and yield the finest flowers." The other engraving, of a pentstemon, decorates the early slogan of the firm: " 'Park,' A Synonym of Dependability—Your SUCCESS and PLEASURE Are More to *Park* Than Your *Money*. Flower Book Sent Free." Today the catalogue is still free and, comparatively speaking, the prices are still low.

All this material—poem, prayer, photographs, and engravings—surrounds a 1959 letter of news and inspiration, which begins, "Dear Patron Friends," and ends, "Floricordially yours, Geo. B. Park, President." It is a gentle, unpushing letter, inviting you to find Peace in your garden and "in the soul, too," before it discusses some of Park's 1959 "exclusives" —Double Gloxinia, Begonia Cinderella, and Park's Cherry Tart. Though they are pretty, I do not happen to like any of these three new flowers as well as I do some of Park's other recent introductions, like the two F_1 hybrid petunias, the lacy wine-red Sugar Plum, and Blue Lace, which is a deep violet-blue with deeper blue veinings, or the flower named Bartonia Stella Polaris. Stella's blossoms are five-pointed golden-yellow stars from the center of which burst long yellow fragrant stamens; the flowers open at sundown, says the catalogue, and remain lovely through half the next day. The bartonia is what Park calls a "ha." This abbreviation, which sounded like an aside, was cleared up in the Key to Symbols on page 20: "ha" in Park terminology is a "hardy annual," a flower whose seeds can be planted outdoors in the spring or fall, while a "half-hardy annual," or "hha," must be started indoors in the spring. We do not make the distinction in the North, where only perennials are called hardy. I suspect that some of the book's horticultural advice may need to be modified for Northern gardens, but it does at least give warnings about annuals particularly susceptible to frost and is informative about the comparative hardiness of its perennials. A great many seeds, old-fashioned or modern, are varieties that all the seedsmen offer, but on almost every page there are names I never heard of, like *Meconopsis baileyii*, a sky-blue perennial poppy from China, and *Incarvillea delavay*, a hardy gloxinia, also from China, discovered there on a mountaintop by a French missionary. Do you, for instance, know *Nolana*, the Chilean bellflower? Or *Amberboa muricata*, Star of the Desert, from Morocco? Or even *Agathea coelestis*, the blue daisy? I did not. These few examples will give you an idea of Park's infinite variety and of its winning habit of telling you where its seeds originated. The pages are peppered with photographs, so that you may see many of the flowers as well as read about them. There is, too, a particularly inviting list of house plants and of the *seeds* of house plants, and much, much more. It is a catalogue that could hold a gardener's attention for weeks.

To a newcomer, "Park's Flower Book" is at first bewildering. Its alphabetical order is only approximate, and in the listings each kind of seed is followed by a mysterious set of letters. Take this entry for a common flower: "Annual Lupines, ha, C, S." By using the Key, you get the translation: "hardy annual, especially valuable for cutting, prefers or will succeed in semi-shade." There are nine other classifications in the Key, but this is just the start of it. Near the end of the book there is a second table of symbols, in the index. The letters, figures, and cabalistic signs in this table designate the height, color, best use, time of bloom, seed-germination period, and culture of every one of the thousands of seeds offered. As you can see, it takes time to find your way around and digest the information in "Park's Book." Your head will swim, your mind boggle at the cataloguer's task, but soon you'll realize that if you do your homework conscientiously, it will not be Park's fault if you do not grow its seeds successfully. Remember? "Your SUCCESS and PLEASURE Are More to *Park* Than Your *Money*."

If the mind is staggered by the detail of the Park catalogue, consider the case of Harry E. Saier, who issues three or four catalogues a year, each of them listing as many as eighteen hundred genera and eighteen thousand kinds of seed. Mr. Saier is not a grower but a collector and distributor of seeds, from the common-garden to the very rare varieties of flowers, vegetables, shrubs, vines, and trees. He does grow a few of the more difficult kinds, but primarily he depends on his two hundred seed collectors, who are stationed all over the world, and on commercial growers from many countries. There is nothing beautiful about his latest catalogue and its hundred and seventy-six pages of small-print lists, interspersed with occasional dim photographs, but it is fascinating to browse in, translating, if you can, the abbreviations made necessary by lack of space. Here is an easy one:

DICRANOSTIGMA (die-kray-no-STIG-ma). Papaveraceae. Plants from Central Asia with orange or yellow fls; best in a light soil—Franchetianum. DICR-1 25¢

Pretty bright fls, 1″ ac; 1 ft.; HB; with us it volunteers every year.

There are Saier branch offices in Australia and New Zealand, and his global lists of clients include nurseries, greenhouses, seedsmen, universities, botanical gardens, and drug manufacturers, but a third of them, he tells me, are amateurs like you and me. When he has time, the indefatigable Mr. Saier also brings out *Saier's Garden Magazine*, published, a note on its cover says, "in junction" with the Saier catalogue. The latest issue is dated January, 1958. It contains a long report by Mr. Saier on a recent trip to Germany and Scotland, a piece titled "D.D.T. and Your Bread," taken from a report to the House of Lords, and articles on the flowering plants of the Kenya Highlands and gardening in New Guinea. Packed in between are odd bits of information, lists of seeds wanted, and editorials by Harry Saier on the inequities of the postal rates, the virtues of stone-ground flour, and the uselessness of giving everyone a college education. (We gardeners must range widely.)

Before I go on to further discoveries, it would be well to discuss some of the old standbys and their 1959 flowers. The major trend this year is, I'm afraid, more than ever toward fussing up and multiplying the number of petals by one means or another. If there is a flower with individuality and a set of delicate single petals, the horticulturists seem determined to double or triple them, to curl them, ruffle them, and, inevitably, to make them bigger. The result is that in many pages of the seed catalogues all blossom form seems to be disappearing, causing one to look fearfully ahead to the time when our garden beds will be full of great shaggy heads, alike except for color, all just great blobs of bloom. For years the emphasis has been on bigger flowers; this year it is on bigger flowers *with fluffier heads*. (Burpee's proudest new aster, for example, which decorates its '59 catalogue cover, is named Curlilocks.) Just as last year, flowers are also being made to imitate other flowers. Burpee's new Miracle Marigolds, the hybrids of its famous Nearest-to-Whites, are described as "big, fully double, pale primrose flowers, of good form in carnation, peony, or chrysanthemum flowered types." No marigold types are mentioned. (Copying peonies appears to be the newest fad; there are also peony *asters* in all the catalogues.) Even Park's beautiful new Giant Double Gloxinias, which the cataloguist terms "a horticultural triumph"—and I have no doubt they are—look to me for all the world like tuberous-rooted begonias. They are lovely, but what I really like in a gloxinia is the pure outlines of its shapely single bell-flowers rising above the broad leaves.

I can't swear that I have put my finger on still another trend, since the words denoting size, like Mammoth, Giant, and Colossal, are still so overwhelming, but it does appear that there is a movement to *decrease* size. Have the customers rebelled? There have always been dwarf plants, of course, but this year it seems to me there are many more of them. In every seed book are Lilliputs, pygmies, midgets, and miniatures. In Breck's, for example, you will find Salvia, Scarlet Pygmy; Ageratum, Midget Blue; Dwarf Fairy Mixture of Candytuft; dwarf French marigolds of the All-America Petite Strain; Little Sweetheart Sweet Peas only eight inches tall. It even offers a collection of midget vegetables. (Schling has some of these, too, including a bean named Tiny Green.) Harris has a marigold called Pygmy Primrose; Park has an aster called the Dwarf Queen and a salpiglossis called the Dwarf Princess. Some might think this new note encouraging, but I find I do not care for dwarfs, midgets, and pygmies a bit more than I do for giants, or, to use this year's term, Super-Giants. Perhaps my unhappiest moment was the discovery in several catalogues of a combination of dwarfism and giantism in the person of "Miss Universe, a brand new idea in Zinnias.

She has stems only two feet high and flowers seven inches across." Poor Miss Universe! She reminds me of my own plight as a girl. I was short, with a great knot of hair at the back of my head. I had not particularly worried about my appearance, though, until an old friend of the family said to me, "I suppose you realize, my dear, that your *hair* is a deformity." Remembering how I felt, I suggest that we do not deform the flowers or, for that matter, dwarf the princesses.

Geophysical entries: In several of the catalogues there is a new petunia called Satellite. I suppose it was inevitable. There is also a calendula named Radar and a gladiolus named Atom. Satellite is quite pretty—a brilliant rose-colored single petunia marked with a perfect white star.

It is encouraging to find Burpee in 1959 somewhat less preoccupied with chromosomes. Last year, everything was "triploid" or "tetraploid," or, in Burpee catalogue language, Tetra—meaning flowers whose chromosome count had been increased by the application of the drug colchicine. This year, the featured snapdragons are not the Tetra Snaps but the new Sentinels—a breed of "tall, erect snapdragons that stand up like sentinels," with as many as forty blossoms on each sturdy spike. Nevertheless, tetraploidism still flourishes in the Burpee lists. Even the unassuming sweet alyssum has not escaped the chemist and his pot, for Burpee now offers Tetra Snowdrift, with heads larger than those of any other annual alyssum. These are nothing, naturally, compared to its new Giant Tetra Zinnias, Shades of Rose, whose blossoms are so big that their portrait in color could hardly be kept within the confines of the page. Indeed, one Burpee flower spilled right out of the book—the new Galaxy Sweet Peas, "a bevy of flowers on one stem." I remain open-minded about Galaxy, though, and shall try some. After all, a bevy of sweet peas conjures up a pretty picture.

The W. Atlee Burpee Company is the world's largest seed house and it carries the endorsement of Luther Burbank on the first page of its catalogue. It is the great experimenter among the seed growers, and I admire it for that, but I trust that its experiments will tend more and more to the perfection of form, as in the Sentinels, which have not a Tetra ruffle on their lips. The fat Burpee catalogue has not changed in appearance in years. You will feel at home in it, and any seeds you order will be good ones.

A gardener who hopes to find the answer to all his needs within one catalogue might be able to, if these needs are not too special, in "Vaughan's Gardening Illustrated," for Vaughan's Seed Company, a Chicago firm, is a nursery as well as a seed house. The "Illustrated" lists a little bit of everything —trees, shrubs, roses, fruits, vines, gladioli, irises, even aquatic plants, as well as the usual run of vegetable and flower seeds. Though there are midgets and giants here, too—and the mixed-up Miss Universe—the emphasis in the seed listings is not on either over- or under-size flowers. Except for an ugly insert in red and green inks, the catalogue is good-looking, and it has more color pages than any other seed book. Vaughan's has another odd distinction: it is very nearly the only company that offers its customers charge accounts. Your money down or nothing doing is the usual rule.

Breck's catalogue, "Better Gardens," should receive an award this year for improvement in appearance. It is very different from last year's book and is now one of the handsomest of the lot. The cover, showing Summer Snow Caladiums and one small gold marigold for contrast, is a lovely study in color and design. The paper of the catalogue has a good gloss, which makes the photographs clearer, and the garden accessories, of which Breck's makes a specialty, no longer distract one's attention from the flower and vegetable seeds. Anyone who can't be bothered with planting many varieties of annual seeds may want to solve his problem with

Breck's Bretton Woods Hotel Flower Garden Mixture. It is one of the best, and a great labor saver. Broadcast in beds, it will give a succession of blooms in harmonious colors all summer long. I ought to know. I still have a family snapshot of two little girls taken against the well-tended beds of the famous old hotel, in the dim and shadowy past, with Mount Washington backing them up, and the flowers are definitely splendid.

Speaking of gardening with less labor, a device called Flowerama burst upon a wondering world a year ago come May. You probably saw the exclamatory full-page ad in the *Times*. A scientific breakthrough! Amazing! Just unroll the Magic Carpet! No topsoil needed! No weeding! No work at all! Hundreds of gorgeous flowers all summer long or your money back! So ran the pretty story. The carpet, it seemed, was a "self-contained medium," and there were "a hundred and one ways to use it!" You could cut it with scissors, place it around trees in circles, make figure eights, grow flowers in the shape of your initials. Although an unbeliever, I unlimbered my scissors and sent off my order for one carpet and my check for $4.95 to Horticulture, Inc., the firm behind the new wonder.

In due course the carpet arrived. It was eighteen feet long, medium gray, and looked like the stuff you put under wall-to-wall carpeting. It also looked a little like the messy interior lining of the soft modern book-mailing envelope. And it was bursting with seed. I cut it in half, thinking to plant one half for a solid mass of bloom and to spell out something with the other half—probably the word WELCOME, though I also considered KEEP OFF. The dry seeds sprang out of the carpet so alarmingly at my first attempt to fashion a letter that I abandoned the idea and decided we would settle for only a nine-foot "riot of color." On a damp, warm evening in early

June, attended by a cloud of mosquitoes, I unrolled the half length of magic carpet and laid it down, as tenderly as if it were a baby, on good soil at the edge of the vegetable garden. Following directions carefully, I watered it. Then nature took over, and we had the wettest June in years, bringing with it a crop of weeds that despite the advertisement's claims did not appear to be in the least cowed by the mulch of Flowerama's carpet.

Our first magic flower was a lone eschscholtzia; it struggled up painfully through the gray fibres in July. By mid-August, there was a small patch of tiny purple-and-brown pea-like blossoms, which I did not recognize. The first week in September gave us a handsome stand of Shirley poppies, about a foot long. A solitary annual chrysanthemum followed. That was all. The carpet had a tendency to scuff up in spite of the continuous wetness, and from June until frost the long blanket lay there looking as if a couple of Boy Scouts had spent the night in our garden, end to end. Eventually our dachshund, who is always inflamed by a stray rag, discovered it and carried bits of the blanket about with him from place to place. Well, considering all, it probably wasn't a fair test, but there can be no doubt that *our* Flowerama was a flop. Perhaps I didn't really want this form of magic. I didn't approve of it, and I notice that plants I don't approve of seldom do well.

The magic carpet we *do* have success with is our vegetable garden. It doesn't come in rolls, and it comes entirely from the Joseph Harris Company, of Moreton Farm, Rochester, New York. (I have found the Harris selective list of flowers equally productive.) The new Harris catalogue is as seductive as ever, with its gleaming black-and-white photographs of vegetables, each one so clear that it might win a prize in a camera contest. I have now decided that we must have at least part of our crop in those new Tendercrop beans, if only because I have just discovered this statement, uttered quietly in small type: "We think this is the best green bean to grow." Having followed similar bits of Harris's forthright advice before, I know that it pays off. I shall also try the new Gold Cup hybrid corn, which Harris says gives astonishing yields, and the new Jade Cross brussels sprout. (This last is listed in several catalogues.) And I shall continue with all my old Harris favorites. Take its Long Season, or Winter Keeper, beets; every beet of this kind that we harvested last fall has kept all winter, buried in a box of sand in the cellar, and we can look forward to borsch well into April from our 1958 beet crop. One of my favorite Harris specialties has been missing in the catalogue for the past couple of years—celery *plants*. (It still lists celery seed.) I wish Harris had not given up growing and selling these plants—especially the variety called Cornell 19, which used to do so well for gardeners of our region, where the season is too short to raise celery from seed.

Speaking of size, Harris offers a corn called Miniature—dwarf plants, with ears five to six inches long, of "supreme quality." I don't know why anyone wants a small ear of corn—not even for convenience in freezing, since corn frozen on the cob is so much less good than corn scored and packed solid. I admit, though, that I am tempted by that Tiny Green bean I spoke of earlier. The Max Schling catalogue describes it as "the nearest thing to Haricot Vert," which, as everyone knows, is the succulent small French green bean. (Schling is another exception to the money-down rule; it will charge orders of over three dollars.)

The catalogues of the famous nurseries of White Flower Farm and Wayside Gardens are too well known to need description, but no garden report would be complete without mention of a few of their new imports. Both houses specialize in importing shrubs, trees, and perennials. White Flower is this year proud of its Symons-Jeune strain of hardy phlox, which comes from England and is said to be more fragrant and to have a longer period of bloom than the average phlox; from England, too, come the Anchusas Opal (light blue) and Pride of Dover (dark sky-blue); from China, though doubtless not directly, come the species syringas, Chinensis (violet) and Yunnanensis (pink), that are the delicate, extra-sweet-smelling ancestors of our lilacs. White Flower also offers a French pussy willow—"not the unreliable wild Pussy Willow," writes Amos Pettingill, "but a fine French cultivated variety." What is unreliable, pray, about the native wild pussies? I have found them trustworthy in every respect. Certainly the ones that grow in our lane have never failed to produce their shimmering buds in the past twenty-five springs, and the branching habit of their twigs is to me far more graceful than that of the large-budded European varieties. Not everything good comes from abroad.

The big news at Wayside this year is perhaps its new line of de Rothschild azaleas, which were developed in England by the late Lionel de Rothschild. They may be ordered only in mixed colors, but the color photographs in the catalogue—four pages of subtly shaded pink, flame, yellow, and gold large-petalled flower clusters, any one of which would look good next to any other—indicate that this is a safe procedure. Wayside's color printing is remarkably true to life. An example is the color picture of its hardy chrysanthemum, Ashes of Roses, which I grew last year. The rosy iridescence of this unusual chrysanthemum is hard to describe, but there it is, glowing on the pages of the catalogue just as it did in our garden from mid-September until frost. Look, too, at the photograph in color of Moonsprite, a charming pale-yellow double floribunda. I grew Moonsprite last year and take issue with the cataloguer's description of this little rose as "chrysanthemum-like." Thank goodness, she is all rose.

Both Wayside and White Flower Farm list an old-fashioned perennial that is an oddity in the plant world. It is *Dictamnus*, commonly known as the gas plant, which secretes with its fragrance a resinous volatile matter that may be ignited on hot summer evenings without damage to the plant. I seem to remember references to the gas plant in my childhood, though I never saw one growing then, and I daresay my grandmother played this little game. I can hardly wait to grow a gas plant of my own and set off its eerie fires.

A panel of five thousand rose growers has selected the new Kordes Perfecta hybrid tea as the Rose of the Year. Jackson & Perkins feature it as such, as do a number of other growers. To my eyes, it is hideous—huge, high-centered, with flashy curled-back petals that are cream-white, dipped at the edges in a harsh carmine. We are told that as the buds open, the carmine spreads and is followed by a suffusion of yellow. I find the prospect unalluring, and greatly prefer J. & P.'s own new rose, a floribunda named Ivory Fashion. The two roses of the past year for me, though, were Charles Mallerin, a hybrid tea developed by the great French hybridist Meilland, and Golden Wings, which the Bosley Nursery, of Mentor, Ohio, introduced. Charles Mallerin, though modern, is not new; its color is the darkest, richest red I know; it is stalwart and fragrant, and in our rose bed it blossomed right up to Thanksgiving. It is not in every catalogue, but Bosley carries it and so does Will Tillotson's Roses, of Watsonville, California, which is where I got mine. Golden Wings is a single rose, with five pale-yellow petals and orange stamens—a charming, airy blossom on a vigorous bush; if left to itself, it will grow to six feet, but it may be kept to average height by judicious pruning. It blooms from early spring until the snow flies and is winter-hardy to the tips. No wonder the American Rose Society awarded it a golden certificate for "performance." I let my two new bushes grow, and they blossomed continuously.

Bosley's catalogue is called "Bosley Rose News"; I commend it to your attention. At first it will appear to be merely a four-page oversize tabloid printed on coated stock, with handsome rose photographs on every page, but in it you will find Bosley's carefully selected list of roses, full of old favorites and interesting new varieties. The current issue shows Golden Wings on the front page. There are four other pictures of new roses, including June Bride, the first white grandiflora, and a remarkable climber named Blossomtime. Along with these are inserted odd items, such as how to make potpourri, and a cheerful paragraph about an Australian doctor who has "a clinical impression" that gardeners have a greater "host-resistance" to cancer than non-gardeners and is doing research to try to confirm this impression.

The most beautiful, and still my favorite of all the rose books, is "Roses of Yesterday and Today," the Will Tillotson catalogue, of which Dorothy C. Stemler is now the editor. A photo of the late Mr. Tillotson forms the frontispiece, and his spirit lingers in the book, together with some of his descriptions of the many old-fashioned and modern roses that the nursery stocks. Mrs. Stemler, it seems to me, has made no missteps. The catalogue is now her own; the tone is a bit less rapturous and the whole presentation is more down to earth, yet still full of charm. The photographs of the roses, which were always hers, are better than ever. The cover this year is Watteau's "L'Amante Inquiète," and it is in delicate color—an innovation. In spite of the title, the Watteau lady looks quite peaceful as she sits holding one rose in her apron. An extra bounty this year is a sepia reproduction on the catalogue's mailing envelope of Peter Bruegel's "Spring Planting," which shows you the planting of a formal garden in the year 1570. I should add that the 1958 Tillotson method of packing dormant roses in polyethylene bags worked perfectly; every rose I ordered from Will Tillotson last spring, including the Charles Mallerin, made it from California to Maine in perfect shape, and lived to thrive, in spite of the change of climate.

Another West Coast catalogue with a distinct personality is "Roy Hennessey's Prize-Winning Roses." Roy Hennessey is Oregon's angry man. A satisfied customer called my attention to his nursery, and I have studied his two latest catalogues. They are straight lists, with descriptions but no illustrations of the many roses that Mr. Hennessey has tested and approves of. (Any rose, or any person, that finds favor in Mr. H.'s eyes is doing all right.) He offers a wide selection of all types—hybrid teas and hybrid perpetuals, tea roses, species roses (these are unusual), mosses and the other shrub roses—including the type he correctly calls "hybrid polyanthas" but which most of us have come to call "floribundas." What he calls "the Floribunda Mess" is one source of Mr. Hennessey's irritability. Another is what he calls "eddication." He is mad at Jackson & Perkins, mad at a Cornell professor named Allen, mad at the American Rose Society, mad at the State of Arizona, and at goodness knows what else.* His diatribes, interspersed throughout the book, may amuse some readers; I found them wearing. Nevertheless, it is obvious that Roy Hennessey loves roses and cares a great deal about developing those that will withstand heat and resist disease. He takes pride in being a nurseryman, not a theorist, or what he calls an "eggspert," yet sometimes he theorizes himself, and not too

* Hennessey refused to send his catalogue to KSW after *The New Yorker* published a piece on Jackson & Perkins, a firm he detested.

clearly. I got completely lost in his discussion of fertilizers, but I was pleased by his reference there to what he calls the "Aegean" stables.

It is soothing to turn from this thorny rose grower to the gentle and scholarly Cecil Houdyshel, who since 1914 has been growing bulbs, tubers, and plants—most of them rare—in La Verne, California. Since 1925 Mr. Houdyshel has been bringing out two brochures a year, titled "Bulbs for Pots and Spring [or Fall] Planting in the Garden." They contain no flower pictures, but as garden literature they are collector's items, and they are important reference books for any serious gardener. I have been lucky enough to acquire a dozen of them, and I wish I had the whole file. The listing of stock is by plant families—Amaryllis, Iris, Lily, Orchid, the Gesneria (gloxinia, African violet, etc.), the Bromeliads, to name but a few—and each brochure by no means gives all the stock available; the emphasis each time is on the grower's present interest. As Mr. Houdyshel describes his wares, he inserts exact and fascinating advice on the culture of a family and even of a special bulb, and there is always a helpful essay on general culture. Each little book leads off with a letter from Mr. Houdyshel, which invariably opens, "Dear Floral Friends." Because his customers really *are* his friends, he sometimes volunteers a bit of personal news—a report on his trip to Carson City in 1958, a visit to San Francisco last summer with his wife. (Mrs. Houdyshel is the African-violet specialist of the firm. The list of named varieties of her violets—plants, and leaves from which plants may be grown—occupies nearly six pages in the latest issue of the brochure.) Mostly, though, Mr. Houdyshel does not like to talk about himself. He is a modest man, and it was not until 1952 that he allowed a snapshot of himself to appear in his catalogue, and then only because he was so proud of his granddaughter and great-granddaughter, who are also in the picture. I am glad he did this, for his kind, thoughtful face makes one understand why his catalogues are so agreeable to read. Most of the letters are philosophical disquisitions on such matters as the creative aspects of gardening, "the pleasures of life," the importance of hobbies, exercise, voting—above all the importance of reading. "I like to quote those ideas as they have been expressed by the great poets or philosophers," he wrote in 1954, and he does quote occasionally, all the way from Juvenal to Thomas Huxley. Sometimes he gives his own horticultural news. "We are happy to announce success in crossing Marica northiana with M. gracilis," he noted offhandedly in 1951. A careful reading of the little books, though, will show that Mr. Houdyshel is a greatly gifted horticulturist, yet he never boasts and he never publishes testimonials. His comments on his triumphantly hardy and floriferous Crinum Cecil Houdyshel are casual, and they finish with this wry comment on how to pronounce his namesake flower: "*Don't* call it who-dee-shell or Hardyshel but as if spelled Howdy-shel." This remarkable man, who is eighty-five years old, is now collecting and working with camellias primarily for his own pleasure, yet I would not be surprised if the results of his work turned up in future catalogues. I am told that he has a hundred varieties already and that he says he hopes to live to be a hundred.

The beginning orchid grower may be interested in J. N. Giridlian, another Californian. His catalogue offers, among many rare bulbs, the easy-to-grow Swan orchids (*Cycnoches chlorochilon*) and the terrestrial orchid Calanthe, which will survive outdoors in temperatures near to zero. For experienced orchid fanciers, there are many kinds of the difficult exotics, including a number of miniature botanical cymbidiums from China and Japan. I was happy to find in the list, along with the orchids and the rare bulbs like the Egret Flower mentioned earlier, five varieties of Dog's Tooth violets and ten of hardy cyclamen. The nursery is called Oakhurst Gardens.

The catalogues of the specialists are often mere listings, but that of Barnhaven Gardens is a charming little book. Lew and Florence Levy, the proprietors of Barnhaven, grow hardy primroses. Their Silver-Dollar Polyanthus, with flowers the size of the coin, are famous, and they also raise Acaulis, Julianas, Asiatic primulas, and many other types. The photographs are lovely and so is the cover, which is a reproduction of a Birket Foster print of a little girl picking wild primroses. The engravings of this Victorian artist appear in all the Barnhaven catalogues and have become a sort of trademark. They are appropriate, since Foster's favorite subjects were English country children picking primroses and cowslips.

Although unillustrated, Miss Silvia Saunders's listings of the rare hybrid peonies developed by her and her late father make a specialty catalogue that can send me off into dreams of garden glory. Here are some of the loveliest aristocrats of the plant world, charmingly named, succinctly described—the work of years by two artist hybridists. Among her best growers, Miss Saunders tells me, are the Lobatas—pink to red shades—and, of the whites, Campagna, Innocence, and the peonies she calls the Early and Late Windflowers. Campagna is a triple hybrid with a green-and-gold center, Innocence is a very tall single white with a greenish center, and the Windflowers are also tall and particularly graceful, with nodding anemone-like flowers. The only true yellows—the Roman Gold and Golden Hind groups—are to be found among the harder-to-grow Luteas, or tree peonies. Anyone who makes the pilgrimage to Wilmington this spring to view Mr. Henry F. du Pont's famous Winterthur Gardens should look for a planting of these yellow Saunders Luteas, which ought to be in bloom there near the end of May.

Richard and Edith Williams, the proprietors of Village Hill Nursery, specialize in herbs, geraniums, and plants for rock gardens, walls, and terraces, and they offer a wide selection of all. Their catalogue is one for the experienced grower, as it does not offer cultural advice, describe the herbs or flowers, or comment on their height or hardiness, but rock gardeners and herb growers will not want to miss it. For a simple, business-like listing of plants, this booklet is well designed and pleasing in its typography.

If it is named varieties of pansies and violas you want, you might try Pitzonka's Pansy Farm and Nursery, Pennsylvania. It sells both pansy seeds and plants by name, which is unusual, as well as in mixtures, and also named varieties of forget-me-nots and English daisies—the flowers that look so good when planted with pansies.

Pitzonka's, though, does not offer my favorite Maggie Mott, a lovely clear-lavender English viola. I used to grow it, but have not been able to find any in a long time. I was therefore delighted to discover Maggie in the last catalogue of the Gardenside Nurseries, of Shelburne, Vermont. This catalogue is really a magazine, *Gardenside Gossip*, which used to be published four times a season but now appears only in the fall. The nursery writes me that the next issue may be the last, because of "diminishing interest in the sort of things we grow." To me this is incredible, even intolerable, because I so *like* the kinds of flowers it grows, and because I have enjoyed this informal sheet so much. (It is done in a printed reproduction of typewriting.) The title is misleading, for here is agreeable, original, useful garden talk, but no gossip. The *Gossip*'s explanation, for instance, of the tetraploid phenomenon is the most illuminating I have read. After discussing how colchicine is used on lily bulbs, it adds quietly that "not all the polyploid plants produced have any value"—

a statement I found refreshing. Lilies are Gardenside's primary product; it has introduced many hardy new varieties, including its prize-winning Pink Trumpets, which are pink hybrids of Regale. The nursery's other special strains are Champlain Hybrids, Shelburne Hybrids, and Barryi. What I most enjoy about these growers, though, is their interest in the long-neglected garden perennials, like betony and painted daisies, and in native New England wild flowers. All through the columns of *Gardenside Gossip* runs a tone of anxiety lest certain flowers they love be lost. They grow wild flowers to preserve them, and they sell them for the same reason, since, as they point out in an earlier issue of *Gossip*, the more that are grown the fewer will be lost forever. Some people, they say, have attacked them on the score that it is wicked and destructive to grow wild flowers commercially. Remembering that years ago we heedlessly let our own cows wipe out the lady's-slippers that grew in our woods, I am pro-Gardenside in this controversy.

The growth of pasture land, the cutting of the forests, and the depredations of flower pickers are responsible for the disappearance of our wild flowers, not the plant growers who collect and propagate them. It seems to me good news, therefore, that Gardenside and another Vermont firm—the Putney Nursery, at Putney—this year list the plants of many scarce wild flowers, like the trilliums, the Virginia bluebell, which is also hardy in New England, and the native orchids, like the fringed and the various lady's-slippers. Even more exciting is to discover that Gardenside sells the *seed* of the fringed gentian, gathered from its own wild stand. This means that the rest of us may buy a packet and scatter the seed in a suitable location, such as a moist, played-out pasture, and thus help preserve this vanishing and lovely flower. On this crusading note of salvation and reclamation, dear floral friends, I leave you.

3

Before the Frost

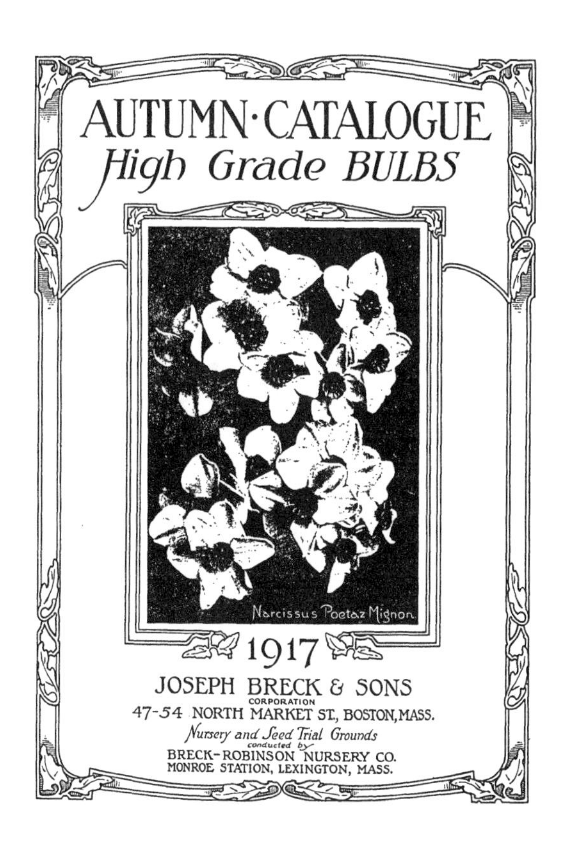

September 26, 1959

It is late August as I begin this report. The days, here in Maine, are noticeably shorter, and lovely summer slips through the hand like Edna Millay's silver fish. The swamp maple flashes its red danger signal, and it's time now for the gardener to make his plans for the house plants that will bring the garden indoors this winter and to order the bulbs that must go into the ground in October and November if the outdoor garden is to come to life again next spring.

Today I'd like nothing more strenuous than to sit still and admire the huge heads of phlox that the wet season has produced in the perennial borders and watch the bees sipping nectar from the poisonous monkshood and plundering the lavender spikes of the veronicas. But a gardener's mind is restless; it runs on ahead, and that is the penalty one pays for the life of culture. Instead of sitting still on the terrace, I go in the house and administer first aid to an ailing African violet—a pink one I count on to bloom for me during the dark winter. I also take a census of the leaves on my four amaryllis bulbs to make sure I am feeding them enough to bring plenty of flower stalks next February and March. Hamilton P. Traub's *Amaryllis Manual* (Macmillan, 1958) tells me that "the achievement of nine leaves will mean that three new inflorescences will be initiated between the bulb scales—one after each third leaf." (What is spring without inflorescence!) I find that one of my amaryllis plants, a last year's salmon Warmenhoven from White Flower Farm, has nine leaves, which is good; my two reds have seven leaves each, which is fair; the fourth, an ancient white, has only five —very poor. Unless I can induce another leaf, that white amaryllis will have only one flowerscape next winter and I will feel shortchanged and somehow derelict. Therefore I move the pot to a slightly sunnier location and give it a dose of "20-20-20 with trace elements." The word "dose" tells the story here; there is no use trying to grow house plants unless one is willing to be a nurse.

For the most part, I raise only the commonest, easiest plants and flowering bulbs. Once in a while I take a flyer and try a difficult exotic. Orchids I abjure entirely, for lack of time. But even with this modest program, I become, in winter, a sort of floor nurse in my spare moments—taking temperatures, rubbing backs, and making a thousand small adjustments when my patients summon me. Here in summer, I recall the anxieties of those midwinter mornings: Were the cyclamens in a strategic place—near, but not touching, a cold windowpane that provided sun but not too much sun? Were the African violets, whose forebears had come from the tall forests of East Africa, getting their filtered forest sunshine and their seventy-degree heat? Was the bottom heat from a radiator under the pots of starting amaryllis enough to encourage growth and still not cook the bulbs? Was it the day to raise the humidity of the plant room by adding water to the pebbled trays? Did the rubber plant need its weekly bath and the sulking clivia miniata its fortnightly shot of chemical fertilizer? Had the mealy bugs moved over from the infested cactuses to the poinsettia? My patients were scattered all over the house, upstairs and down, and I had a ward of many pots in the plant room, which is really part of the kitchen. Some of these ward patients were there because I had raised them or sought them out; some, like the Christmas and orchid cactuses, were there because they were aging members of the family whom sentiment forbade me to abandon; others were accident cases—that is, unexpected gifts, always welcome but sometimes hard to find space for. My floor duty started before breakfast, when I watered the plants in the upstairs rooms, and ended at bedtime, when I checked night temperatures and adjusted the thermostat and the radiators.

Baths could not be worked in until midafternoon, when the kitchen sink was free of cooking utensils. Pillow plumping and words of cheer came at any old time as I passed through a room and paused just long enough to remove a withering leaf from a geranium or loosen the earth in a packed pot or move a plant to a window where it would catch a few rays of the pale winter light.

It is an agreeable life, but it calls for dedication. There was one morning, I remember, when I realized that the temperature I ought to be taking was my husband's; a tendril of the grape ivy on the piano had reached out, as he walked by, and seized him lightly by the wrist. It is all too easy for house plants to take over a home, leaving no room for the owner. Already I am worrying about where to put the pots that must soon be moved in from the terrace—plants like White Flower Farm's lovely trailing lantanas (I have the lavender species, Selloviana, and the pale-yellow hybrid, Cream Carpet). And I have a mammoth six-year-old strelitzia from the Park Seed Company, in its huge tub; where am I going to put *that*? And my loyal old rhizomatous begonias, which have grown so portly—where do *they* fit in? Then there are the terrace fuchsias, but these I can defoliate and store in the cellar. They have done well this summer because they are out of the wind and because our Maine nights are cool enough for them to set buds—that is, temperatures that are often below sixty-five degrees. As for the lantanas, I am eager to try them indoors in hanging baskets; if I can keep them warm enough at night (a minimum of sixty degrees) and give them five hours of sunshine, they will bloom (or they are supposed to) every day of the year.

These knowing remarks about minimum temperatures and required hours of sunlight do not come from my own store of knowledge; they come from Ernesta Drinker Ballard's *Garden in Your House* (Harper, 1958), and I make them not to show off but to prove what a useful book it is. All gardeners need reference books, especially indoor gardeners. Outdoors, nature is apt to take over and save you from many a stupidity, but indoors you are strictly on your own. Manuals like the one I quoted earlier on the amaryllis are useful, and Dr. Traub's is probably the best ever written on this subject, yet his sentence about inflorescences has more significance to me than the facts it conveys, for his style and his rather private terminology of the specialist are characteristic of one whole school of garden writing. Another manual I often consult is a guide to the gesneriads, by Harold E. Moore, Jr., a professor of botany at Cornell. It is *African Violets, Gloxinias, and Their Relatives* (Macmillan, 1957), and it, too, is not easy, even though, for its type, it is a particularly attractive book, what with its five-color plates and Professor Moore's occasional digressions from the textbook style when he describes his journeys in search of plants. But if the scientists' books do not make the most pleasurable sort of reading, I greatly prefer them to the common run of garden books written by the cozy ladies and punning gentlemen of the Green Thumb school, who have brought gardeners and gardening into disrepute by their sentimentality, their ecstatic or facetious prose, and their inexact advice, which has killed off many a plant and many a would-be plant-grower. There are always some garden books that belong to neither of these schools, for they are well written, accurate, and actually fun to read. Because they are as rare as the plants they sometimes describe, I might mention a few I have particularly enjoyed in the past year.

The best book on house plants I've ever come across is the one I stole my temperature facts from—*Garden in Your House*. It is a handsome, glossy volume, with nearly a hundred photographs, which picture such a luscious growth of healthy and unusual plants that at first they filled this amateur gardener with despair. But Mrs. Ballard puts on no airs. Though she knows her botany as well as her horticulture, she is a clear and concise writer who never speaks through the dark glass of technical language. She is also a practitioner, not a theorist, and has herself grown every one of the more than five hundred and fifty varieties of plants she lists, in dictionary form, in the section of her book titled "Selection of Plants." Each plant is dealt with in a brief paragraph giving its origins and oddities, its merits and demerits, as well as the exact directions for making it thrive. Her first chapter, on the general culture of indoor plants, is excellent —condensed but all-inclusive. The second chapter describes "Some Indoor Gardens," with imaginative suggestions for what to grow where. The final chapter takes care of practical matters, such as formulas for potting mixtures, tables of minimum night temperatures and required hours of sunlight, and lists of the sources of supply for plants and indoor-gardening materials.

Mrs. Ballard's interest is obviously the unusual plant, and I sense that she is able to take to her heart a plant just because it *is* unusual, which is more than I can do. She also entirely omits the common garden bulbs that most of us like to grow indoors for color or scent. But no one who has read her book is ever again going to be content merely to grow the old reliables like geraniums, begonias, and ivies, or to force a few paper-white narcissi for winter bloom. I shall never desert the bulbs, though, and last winter I think I got more pleasure from a pot of February Gold daffodils than from anything else I raised, unless it was my pots of freesias. February Gold, which is a medium-small, all-yellow narcissus of the cyclamen type, for me proved to be January Gold; it opened its first flowers on New Year's Day. That was the miracle. There is no trick to growing it in pots if one has a cool cellar, and Wayside Gardens, where I got my bulbs, says it can also be grown in bowls, like the paper-whites.

If Mrs. Ballard omits the garden bulbs, Elizabeth Lawrence's *The Little Bulbs, a Tale of Two Gardens* (Criterion, 1957) really does them justice. It is a delightful book, and reading it is quite as much a literary pleasure as a gardening one. Miss Lawrence, too, is a botanist and horticulturist; still better, she is a writer with a sense of humor and a sense of beauty. The smallest of the flowering bulbs are her favorites. She has a chapter on growing them in pots, but she is primarily an outdoor gardener. The two gardens of her title are a Mr. Krippendorf's, in Ohio, and her own, in North Carolina. "Mine is a small city back yard laid out in flower beds and gravel walks," she tells us, "with a scrap of pine woods in the background; Mr. Krippendorf's is hundreds of acres of virgin forest. . . . While I content myself with a single daffodil, Mr. Krippendorf likes to see ten thousand at a glance. Where I plant half a dozen, he plants a hundred." We are given a glimpse of the Ohio gardener "in his leaf-colored jacket, with a red bandanna around his neck," but, with the formality of the Southerner, Miss Lawrence never refers to him by his full name.

As soon as spring is in the air Mr. Krippendorf and I begin an antiphonal chorus, like two frogs in neighboring ponds: What have you in bloom, I ask, and he answers from Ohio that there are hellebores in the woods, and crocuses and snowdrops and winter aconite. Then I tell him that in North Carolina the early daffodils are out but that the aconites are gone and the crocuses past their best.

This exchange of intelligence between the two friends, with its counterpoint of North and South, is amusing and informative, but Miss Lawrence has many other garden correspondents whom we get to know before the book is over, and she is well acquainted with many gardens besides her own and Mr. Krippendorf's, including those in books. These last, she says, are often as real to her as the gardens she has seen. We come to know them, too, and their authors. I am grateful to this book not only for all it has told me about the fragile botanical tulips, the wood sorrels, the hardy cyclamens, the squills and little daffodils, the snowdrops and snowflakes, and most of the other small bulbs, but for introducing me to the books on gardens that Miss Lawrence values.

One of these books is by "Mrs. London," who gardened and wrote books on gardening in Bayswater, London, more than a century ago. Like Mr. Krippendorf, she is known to us by no fuller name. I have never seen Mrs. Loudon's magnum opus, the *Ladies' Flower Garden* trilogy on ornamental annuals, bulbous plants, and perennials, illustrated in color, of which the second book is *The Ladies' Flower Garden of Ornamental Bulbous Plants*, a volume Miss Lawrence says she still constantly consults. I do, though, own another book by Mrs. Loudon, which came down to me among the family books, and I have been reading it all the past year with great profit and pleasure. Mine is the first American edition (Wiley & Putnam, New York, 1846) of what I take to be two of Mrs. Loudon's earliest works—*Gardening for Ladies* and *The Ladies' Companion to the Flower-Garden*—combined into one volume and illustrated by the original engravings. These works had apparently been popular in England for some years before they were published here, because the American editions are reprinted from the revised third English editions, the prefaces for which Mrs. Loudon signed in 1840 and 1842. The editor of the American book, A. J. Downing—a great man for misplaced commas—announces blandly in *his* preface that he has cut the portions relating to kitchen gardening and says that Mrs. Loudon's books are intended especially for "lady gardeners,—a class of amateurs which, in England, numbers many and zealous devotees, even among the highest ranks. It is to be hoped, that the dissemination in this country of works like the present volume may increase among our own fair countrywomen, the tastes for these delightful occupations in the open air, which are so conducive to their own health and to the beauty and interest of our homes." I agree that our homes might indeed be more interesting if we took some of the suggestions Mrs. Loudon makes in *The Companion*, which is an encyclopedia of everything that pertains to the indoor and outdoor garden. Consider, for example, her remarks on the "Pleasure-Ground," which she defines as that portion of a country residence not devoted exclusively to utility or profit. The entry occupies more than a page of fine print, and I quote at length because of the vistas and possibilities it opens:

In former times, when the geometrical style of laying out grounds prevailed, a pleasure-ground consisted of terrace walks, a bowling green, a labyrinth, a bosquet, a small wood, a shady walk commonly of nut trees, but sometimes a shady avenue, with ponds of water, fountains, statues, & c. In modern times the pleasure-ground consists chiefly of a lawn of smoothly-shaven turf, interspersed with beds of flowers, groups of shrubs, scattered trees, and, according to circumstances, with part or the whole of the scenes and objects which belong to a pleasure-ground in the ancient style. . . . Walks may stretch away . . . to shrubbery, which . . . is commonly framed into an Arboretum and Fruticetum . . . and there may be rustic structures such as woodhouses, mosshouses, roothouses, rockhouses, or cyclopæan cottages, Swiss cottages, common covered seats, exposed seats of wood and stone, temples, ruins, grottoes, caverns, imitations of ancient buildings.

She goes on to suggest a visit to

Alton Towers, in Staffordshire, where, in addition to the objects mentioned, may be seen pagodas, hermitages, an imitation of Stonehenge, and of other Druidical monuments, shellwork, gilt domes, and huge blocks of massy rock, bridges, viaducts, and many other curious objects.

It is not often that Mrs. Loudon takes off on such a spree. Most of *The Companion* is a dictionary of the English and botanic names of "the most popular flowers," with directions for their culture, which are in many cases as useful today as they ever were. She even includes a few *un*popular flowers, the most invigorating of which I think is

Momordica—*Cucurbitaceæ*—The Squirting Cucumber. An annual gourd-like plant, with woolly leaves, and yellow flowers, the fruit of which resembles a small cucumber; and which, when ripe, bursts the moment it is touched, scattering its seeds, and the half-liquid, pulpy matter in which they are contained, to a considerable distance. This quality made it a favourite, in gardens, a century ago, when some people were yet in a state of sufficient barbarism to find amusement in the annoyance of others; but it has now deservedly fallen into disrepute, and is seldom grown.

I'm afraid I know several little modern American barbarians who would be delighted if their grandmothers would grow them a supply of squirting cucumbers.

It would be wrong to give the impression that Mrs. Loudon's books are merely quaint. They are prime examples of what garden-book writing should be—unsentimental, thorough, readable, and scientific. *Gardening for Ladies*, after admonishing the ladies to wear long gauntlets for the protection of their hands and arms, tells them how to dig and manure the soil, how to sow seeds, plant bulbs, transplant seedlings, prune, propagate, and graft (the grafting section is particularly detailed and well illustrated), and how to garden in a window or a small greenhouse. She delights me by saying that the common unglazed red earthenware flower-pots are decidedly the best, a judgment in which the up-to-date Mrs. Ballard concurs, adding that she has the most satisfactory results with the old-fashioned pots. So do I, but try and find them! Every plant one buys nowadays comes in a hideous green, lavender, or gray plastic pot, often striated with pink. If pots like these were merely hideous, they might be forgiven, but they are worse than hideous, they are non-porous. The soil cannot breathe, it does not drain properly, and the danger of overwatering is great. My greenhouse man tells me that the trade likes plastic pots chiefly because they save money; there is less breakage, and shipping costs are less because the pots are lighter. To be sure, plastic does not encourage fungus growths or accumulate the salts from fertilizers that rise to the tops of clay pots, sometimes killing a drooping leaf or two, but these dangers are far less serious than the danger of bad drainage and overwatering. I have tried, but I cannot seem to manage most plants in plastic, and now that I have begged or bought all the old-fashioned pots in my neighborhood, I am stuck. It seems to me there would be a brisk business for anyone almost anywhere who would stock the red earthenware kind in all sizes. How pretty they are, too! Are we really to lose them entirely?*

Often the best source for the commoner house plants is the nearest greenhouse or flower shop; for unusual ones, one must send away. I find that I am most apt to send away to

* Earthenware pots are available today. Florists and growers, however, often prefer the plastic ones—they are lighter and easier to ship.

the firms that never fail to mail me their catalogues without my having to write and ask for them. I mention this only because catalogues that do not arrive without prodding are one of the annoyances of gardening. One can't expect a catalogue if one has not bought anything within the year, but a customer who buys does expect it.

The catalogues of the Geo. W. Park Seed Company, of Greenwood, South Carolina, do come along regularly, and they are an excellent source for a limited but interesting selection of house plants and for a far from limited list of the seeds of house plants. Park packages its plants with care, and the ones that made the journey from South Carolina to me in Maine arrived in perfect shape, except for one that was delayed in the mails, and this Park promptly replaced. The 1959 fall catalogue offers Park's unusual double gloxinias, and if you like the showy doubles, you will be glad to hear that two blues have been added to last spring's two scarlets. Park also offers the tubers of many singles for Christmas or midwinter bloom. For anyone who has a cool place to store a gloxinia during its rest period, it is one of the easiest of house plants, and it seems to last forever. Right now I have in bloom a pure-white single that is well over ten years old, and though I have not bothered to repot it in years, which is against all the rules, it has produced a spectacular show this summer—more than thirty blossoms out at one time, followed by at least twenty others. (Of course, I fed it heavily to make up for not giving it new soil.) I mention breaking the rules only to show that, important as the reference books are, it isn't necessary to follow them slavishly. Besides, horticulturists do not always agree. For example, White Flower Farm, in the directions it sends with its clivia-miniata plants, issues solemn warnings against letting a ray of sun touch the leaves. I obeyed the instructions and in two years have raised leaf after leaf, but not a bloom. I notice that Mrs. Ballard recommends two hours of winter sunshine a day for clivia and partial shade outdoors in summer, so I have set my plant out under a shrub, and I should soon know which rule to follow. White Flower offers to replace the plants that fail to bloom, but perhaps the only replacement should be a new set of directions.

To return to Park's catalogue, the company is offering a new hybrid fibrous-rooted pale-pink begonia with the unhappy name of Blushing Baby. The infant looks pretty in its picture, though, in spite of its embarrassment. Another Park specialty, and a plant not always easy to come by, is *Crossandra undulaefolia*, or, by its newly corrected name, *Crossandra infundibuliformis*. Take your choice between the tongue twisters, but before you decide to buy this handsome evergreen, nearly ever-blooming plant, with its glossy dark-green leaves and orange-pink clusters of florets, make sure you have time to pamper it. (Again you hear Mrs. Ballard speaking; she calls Crossandra "exacting.")

One of the most pleasing catalogues in which to search for house plants is put out by Ervin and Mary Ellen Ross, of Merry Gardens, Camden, Maine. The Rosses specialize in geraniums, begonias, and rare house plants, and also grow some herbs and perennials. They are a young couple who must have labored hard to accumulate and propagate all the unusual plants they offer in the thirty pages of their good-looking but unpretentious little book. The paper is coated, the typography is good, and on nearly every page there are from three to six small black-and-white photographs of plants. These pictures are so clear that I believe I have at last learned from one of them the name of my ancient rhizomatous begonias. A neighbor gave me a medium-sized plant of them years ago, but she did not know the name of the variety, nor did I. Now I am quite sure that it must be Leo C. Shippy: "Leaves bronzy-emerald-green with red veins, red hairs underside, giving ruffled effect to the margins; pink blooms." These blooms, growing on tall, delicate stems, start as deep-pink buds and turn to pale-pink fluffy clusters of flowers, which blossom ceaselessly through the four darkest months. My one plant has become two huge ones over the years, even though I keep breaking off pieces of the root stock and giving them away. Leo Shippy has been an easy patient and his descendants people our town, but perhaps I should add that not everyone loves him. We once had a practical nurse in the house. She was a friendly, agreeable woman, but every time she passed a pot of these begonias she shuddered and said, "I *hate* that plant!" For a time, it was nip and tuck whether I could keep both Leo and the nurse.

The Merry Gardens catalogue starts off with eight pages of named varieties of geraniums, which are classified into groups—the miniatures, two to three inches tall; the fancy-leaved and the scented-leaved; the odd and rare; the rosebud, ivy, and Lady Washington types; and many more. Next come five pages of begonias—the fibrous-rooted; the rhizomatous; the angel wings; the hairy-leaved; the small-leaved branching, and so on. (No tuberous-rooted kinds, though. I don't know whether this is because tuberous begonias bloom only in summer or because the Rosses, like me, have no real affection for these opulent plants, with their big, lolling flowers.) The list continues with foliage plants, flowering plants—everything from the common Patient Lucy to the rare variegated marica—then vines, cactuses and other succulents (including eight varieties of the easy-to-grow kalanchoe), mosses, and ferns. A very special list is of ivies, which contains some real rarities. If one had the space, it would be fun to try them all—the small- and medium-leaved, the pleated-, heart-, and fan-leaved, the crinkle-edged, the lacy Ivalace and the marbled California Gold, and Green Feather, which has the smallest leaves of all. The Rosses are not the sort of cataloguers who tout their wares or turn lyrical in their language. The furthest they will go is to add "Rare" or "Choice" to the description of a plant, or to make a mild comment like "Nice climber." They indicate, by groupings, the plants that do best in full sun or in semi-shade, but otherwise they do not attempt to give any cultural directions.

One of the great charms of the little brochures that are issued twice a year as a continuing catalogue by Cecil Houdyshel is their exact and interesting cultural directions. This autumn's issue of "Bulbs for Pots and Fall Planting in the Garden" starts, as usual, with Mr. Houdyshel's personal letter and his general essay on how to grow bulbs. For each bulb or plant that needs special culture, instructions are given with the descriptions of the plants, as well as informal remarks such as "There is only one reason you do not all buy Nerines. You do not know their beauty and ease of culture." Nerines are of the amaryllis family, and Mr. Houdyshel offers seven varieties. His listing of amaryllids runs from Agapanthus, the Blue Lily of the Nile, to Zephyranthes, the Fairy Lily, the smallest of this family. What a rich list this is! Consider that it contains eighteen named varieties of the big Dutch hybrid amaryllis from the famous Warmenhoven and Ludwig strains, and at prices that are often lower than the unnamed varieties of these strains offered in the East. I cannot begin to mention all the exciting rare plants that turn up in this small booklet. The section on the Gesneria family lists a dozen espiscias and a half-dozen columneas (these are epiphytic vines with vivid-red flowers that may be grown in pots or baskets in a sunny window), and under the heading of African violets there are sixty-eight named varieties of plants, not to speak of leaves from a couple of hundred more, for home propagation. The violets are Mrs. Houdyshel's province. I suspect that hers is a very distinguished selection indeed, but the African-violet world is one of such complexity and such specialization that I cannot pretend to be sure. I *can* state firmly, though, that when I asked her to choose for me a single-flowered pale lavender, a most enchanting specimen turned up by air mail, fresh as—well, a daisy. Its name is Santa Maria, and, as the catalogue says, it has "very pretty dark quilted leaves" and is a "powder-blue ruffled single." A new variety on the list this fall is Ember Dream, "bright raspberry red semi-dbl. on dark green quilted fol." (A plant of it, too, made the long journey successfully, and by parcel post.)

Mrs. Houdyshel's descriptions are straightforward, compared with the descriptions of African violets in many catalogues, which sometimes employ so special a terminology that unless one is a fancier, one is lost. I know that flowers can be single, double, and fringed. I know dimly that there are "boy" leaves (plain dark-green blade) and "girl" leaves (pale spot at base of leaf); that there are ruffled, fringed, wavy, fluted, quilted, spooned, and scalloped leaves; and that there are leaves of the Supreme, Amazon, and Dupont types, but I don't know what these types are except that their leaves are big. I am probably the only woman in the United States who has happily grown African violets for twenty years (often, despite vicissitudes, with considerable success) but who has been unwilling to learn African-violet double-talk or follow the usual stringent instructions for growing the plants. For one thing, I often let mine expand, instead of cutting out the side shoots and holding the plants to a single crown, as the books advise. They flower profusely even so. Why not? No one holds violets to a single crown in their native habitat. (Of course, eventually they do outgrow their pots and have to be divided.) The early books on African violets all warned against putting them in the sun and watering them from the top, but these instructions, too, were against nature and have been abandoned.

As almost everyone knows, the African violet is not a

violet at all, not even a cousin of the violet. It is a saintpaulia, named for and discovered in East Africa by Baron Walter von Saint Paul. Professor Moore gives the Baron's full name—Adalbert Emil Walter Redcliffe Le Tanneux von Saint Paul-Illaire—which fits right in with the rest of the African-violet complications. I was not at all surprised, then, to discover in the Professor's book that our modern African saintpaulias descend chiefly from two species, *ionantha* and *confusa.* Confusa is right!

Despite all the nonsense, saintpaulias are exquisite. To me they are like wild flowers in the house, almost as delicate as a bowlful of hepaticas or wood violets. Two more good sources for them are the catalogues issued by Tinari Greenhouses (Bethayres, Pennsylvania) and Buell's (Eastford, Connecticut). The Tinari booklet has the advantage of showing a good many of the flowers in color, which is a help in selection. Buell's supplement to its 1958–59 catalogue offers a long list of named varieties but no pictures. Buell's also specializes in gloxinia hybrids, but if you wish to buy these by name and color, you should ask for the complete catalogue. The name "gloxinia," by the way, is another misnomer. The ones commonly grown as house plants are really sinningia hybrids of the Fyfiana group, so called because a Scottish gardener named Fyfe made a happy cross of the already misnamed gloxinia way back in 1845. Buell's has been growing and crossing the large-flowered hybrids for more than sixteen years and offers them as plants, tubers, or seed. The catalogue also has a list of rare sinningias, including the "slipper-type" gloxinias, which are interesting to grow and hard to find.

Though, as I've said, I'm not too fond of tuberous begonias, I am bound to report that the catalogue of Vetterle & Reinelt, the firm that has developed the best strains of the variety, is one of the most beautiful I've seen, and its illustrations, in both color and black-and-white, are remarkable examples of flower photography. Send for it anyway, if only to see in melting colors the huge blossoms—the rose forms, the camellias, the ruffled camellias, and, loveliest of all, the rose picotees. My eye admires them even while my taste rejects them, which I think is in part because I live in an old New England farmhouse, against which these lush blooms look out of period. For any modern house, they have perhaps just the right amount of flamboyance to contrast with the geometric lines of the architecture. V. & R. also offers the small-flowered hanging-basket begonias, which look good anywhere, and the new hybrid polyantha primroses, which are one of its specialties. The firm is most famous, though, for originating the large-flowered Pacific Strain hybrid delphiniums, which nearly all gardeners grow and nearly all nurseries have for sale. Through this catalogue, you may buy them direct from the originators, either as seed, to start in fall or in spring, or as seedlings, which are sold only by the dozen, for September and October or March and April delivery. In the color pages you may see the subtle shades of these most decorative of border flowers. They are sold by series rather than by varieties, the blue strains with descriptive names like Summer Skies, Blue Bird, and Blue Jay; the whites, which are Galahad and Percival; the lavender and the new pinks, also with names from *The Idylls of the King*—Lancelot, Guinevere, King Arthur, and other familiars of the Round Table.

Now for a few notes on spring bulbs for autumn planting outdoors. Everybody has a favorite source for bulbs, and because practically every nursery and seed house in the country offers the traditional tulips, daffodils, and other flowering bulbs from Holland, I cannot begin to cover the field. I am pleased to note, though, that Burpee, whose catalogues have usually been distinguished for the variety and quality of the stock they offer but not for beauty, has now brought out a really handsome twenty-page catalogue of bulbs for fall planting, printed on glossy stock and illustrated mostly in color—and very good color it is. The front cover shows four giant trumpet daffodils in a pleasing pattern and the back cover five giant tulips. These giants are in the Burpee tradition (big, big, big), but all through the book are smaller types as well, though not a very wide selection.

Wayside Gardens offers a far wider choice in the bulb section of its fall catalogue for 1959. The book, which covers more than bulbs, is well groomed and good-looking, as always, but many of its color pictures are repeats from the 1957 fall catalogue. Wayside's bulbs come from the firm of Grullemans, in Lisse, Holland, and I can testify to their high quality. I was particularly pleased with the botanical tulips I planted last fall, for, despite a rough winter, these delightful small flowers showed up in force in the spring—the marjolettis, the strange acuminatas, the saxatiles, and so on. This year, Wayside features a new bearded tulip called the Skipper, as does the Park Seed Company fall catalogue. Oddly, Park credits Hans Grullemans, of Lisse, Holland, as the originator, but Wayside does not, even though (or perhaps because) Wayside's president, Mr. J. J. Grullemans, is a member of the Grullemans firm in Lisse. The Skipper is a huge, goblet-shaped tulip with bluish-violet petals shading to a tawny brown on their fringed edges—a true original in the tulip family.

Park also makes much of what it calls "bouquet tulips" and lists twelve kinds. (Wayside has three, but calls them "multi-flowered.") In Park's color pictures and black-and-white photographs they look charming—a cluster of four or five tulips from one bulb, growing on upright branches from a central stalk, and the colors are subtle and varied. It is an easy way to get a show with a small number of bulbs, but until I try some, and I plan to, I can't be sure that I shall like them as well as the single-flowered tulips on their firm, graceful stems.

Of all the bulb catalogues I have at hand, the most copious, and the most comely, to *my* eye, is that of P. de Jager & Sons, of South Hamilton, Massachusetts, whose nurseries and head office are at Heiloo, Holland. Their 1959 book is bigger than ever before—a hundred pages, often with twenty varieties to a page, which will give you an idea of the many choices you will be faced with. Nevertheless, the pages are comfortably moderate in size, and the presentation is so orderly, the typography is so clear, and the plentiful photographs in color and black-and-white are so good to look at that the fat little book has the charm of a neat, homelike interior in a Dutch painting. One can spend many pleasant hours of selection or of dream inside its covers. Unlike the standard catalogue pictures of flowering bulbs, the sixteen color pages, most of them carrying four separate chastely beautiful portraits of a single tulip, narcissus, or daffodil flower, never commit the common mistake of making a big-flowered type seem so huge as to be monstrous. The text is equally restrained, without attempts to overwrite a description or to overpraise, yet the catalogue writer does allow himself to feel enthusiastic when he feels so, which is most of the time (". . . a striking flower of good poise, lovely for the garden, vigorous grower"). Also, the varieties the firm considers outstanding are listed in larger type. "We make a great specialty of tulips," says the book—certainly an understatement, for there are thirty-six pages of tulips of eighteen distinct types. I tried to count the varieties, but lost track somewhere past five hundred. There are perhaps half as many daffodils and narcissi, and of the loveliest sorts. If you have read Miss Lawrence on the little daffodils and wish to try some, you will here find many of the ones she speaks of—for instance, the Angel's Tears daffodil (*Narcissus triandrus alba*) and the rarer Hawera and April Tears. After the narcissi come all the other bulbs, from the alliums (the ornamental garlics) to the veltheimias. A lot of them are fairly unusual—tecophilea, the Chilean blue crocus; Alpine hyacinth (*Hyacinthus amethystinus*); lachenalia, or Cape cowslip; puschkinia, the Lebanon squill. Cape cowslips, de Jager points out, may be grown indoors in pots or bowls. The catalogue is careful to tell what bulbs may be so grown, and thus it is of great interest to the house-plant grower as well as to the outdoor gardener. Most of the bulbs are sold only by the dozen, but some rare (and therefore expensive) ones may be bought singly. If you wish the new daffodil called Air Marshal (large flower, broad perianth of intense golden yellow, straight cup of orange-red with a frilled mouth), it will cost you six dollars a bulb, but this price is unusual. Many of the new varieties sell for as little as thirty-five cents, and the bulbs that sell by the dozen are, on the whole, moderately priced.

Lilies are perhaps the most photogenic of all flowers—far more so than roses, say—and lily catalogues are apt to be riotously gay when they are done in color. To some people, apparently, lilies still mean only Easter lilies and the other old-fashioned formal or mournful or churchly whites, but this must be because they haven't seen the modern lilies in all their wild and glorious colors. Some people, too, think lily bulbs are hard to grow; actually, they are easy. The only trouble I've ever had with them was my own fault. Last fall a late order and an early freeze did me in; when my bulbs arrived from Oregon, the ground was cast iron, and it remained that way until spring. We waited hopefully for a thaw, which never came; we applied mulches of straw and fresh manure, but were never able to soften the soil. Finally I got desperate, and my neighbors, had they only known, might have witnessed the comic spectacle of me and my eager family attacking a flower border with a crowbar and dropping bulbs into the icy rubble that resulted. Of course, not a bulb came up in the spring. If I had had Jan de Graaff's *New Book of Lilies* (Barrows, 1951) then, or if I had used common sense, I would have known enough to plant the bulbs in pots and put them in a cool but frost-free place. Mr. de Graaff's chapter on growing lilies indoors indicates that if I had, I would have had flowers by March.

Jan de Graaff's is a name to conjure with, for he is equally famous as hybridizer and as grower of lilies at his Oregon Bulb Farms, in Gresham, Oregon. He does not sell at retail, but a few weeks ago I was lucky enough to see the handsome catalogue and color sheet that he gets out for the wholesale trade. Shortly thereafter I discovered an announcement in White Flower Farm's *Notes* that it is now offering the de Graaff line of bulbs, and, to my delight, the identical color sheet was enclosed. (This was White Flower's first venture into color, by the way.) By subscribing to the bi-monthly *White Flower Farm Notes* (Litchfield, Connecticut) you may decide for yourself which of the twenty-one de Graaff lilies on the color sheet would best suit your taste and your garden. There are Heart's Desire and the other new strains of the famous Olympic white lilies; the Orientals, like rosy Jillian Wallace and the two showy banded auratums; the gay Fiesta and Bellingham hybrids; the informal Mid-Century and Sunburst strains; the new tawny Regal lily, Royal Gold, and four other new trumpets—Black Dragon, Golden Clarion, Emerald, and Pink Perfection. I like every one of these lilies except Pink

Perfection, which looks too much like raspberry ice cream, and the red-banded auratum, which looks like a girl with too much rouge. Particularly lovely is the picture of the airy wild lily Lankongense, shown in pink; its colors can range all the way from orange to lacquer red.

Finally, I must reluctantly report that the newest fad in growing plants indoors is to grow them under lights. *Gardening Indoors Under Lights*, by Frederick H. and Jacqueline L. Kranz (Viking, 1957), tells you exactly how to do it. The whole idea bores me, but if you feel you must create artificial sunlight for propagating flowers and vegetables in the house, here are all the answers. I am disinclined to fuss with wiring, fluorescent tubes, automatic time switches, and tables of light intensities, and I find that the chief pleasure of growing things indoors is that it can be a natural process—a simple way to bring nature into the house. Though I'm willing to be a floor nurse, I have no intention of becoming an electrician. The furthest I will go is to move a budding amaryllis under a reading lamp on a dark day. I shall let somebody else worry about the asbestos-covered wires and about whether the bashful African violet is improperly exposed under the strong Fluoro-Al light.

4

The Changing Rose, the Enduring Cabbage

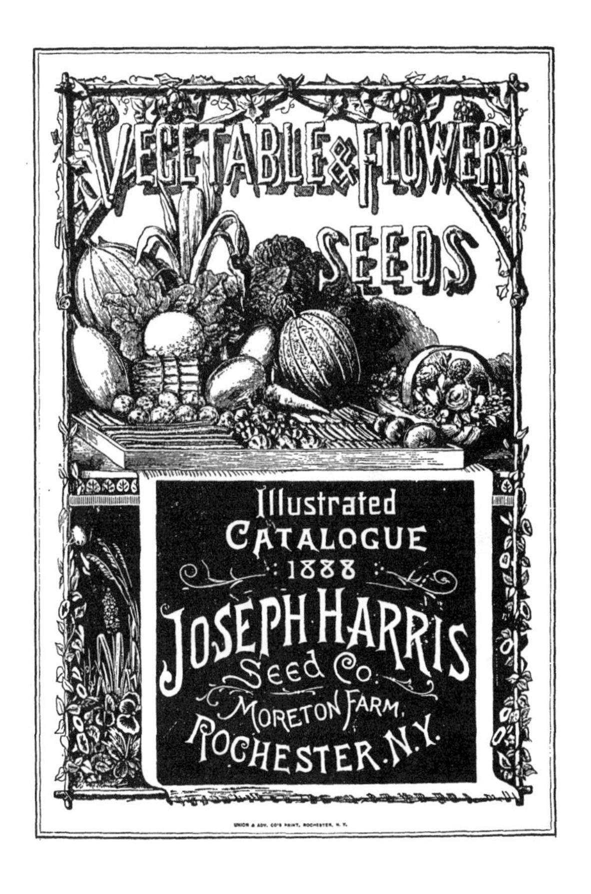

March 5, 1960

In this cold winter of 1960—cold, that is, in the Maine coastal town from which I write—I have been thinking about the fragrance of flowers, a subject that has occupied my mind and nose all year, but one that the garden-catalogue writers, except for the rose growers, tend to neglect. Even the rose men, because they are stuck with a great many scentless varieties, do not give it the emphasis they might. My own nose is not a very good organ, because I am a heavy smoker, but nevertheless I value fragrance and find it one of the charms of a garden, whether indoors or out. The flowers I enjoy most that are in bloom indoors just now are my two big pots of freesias, a white and a yellow; their delicate scent is there for the sniffing, but it does not overwhelm the room, as some lilies do. Colette once wrote that the ideal place for the lily is the kitchen garden, and remembers that in the garden of her childhood "it was lord of all it surveyed by virtue of its scent and its striking appearance," but she goes on to say that her mother would sometimes call from her chair, "Close the garden gate a little, the lilies are making the drawing room uninhabitable!" The young Colette was allowed to pick these lilies by the armful and

> . . . bear them away to place on the altar for the Hail Mary celebration. The church was hot and stuffy and the children burdened with flowers. The uncompromising scent of the lilies made the air dense and disconcerted the hymn singing. Some of the congregation hurriedly left the building, while others let their heads droop and slumbered, transported by a strange drowsiness.

We all know this phenomenon—the somnolent Easter Day service in a lily-laden church. Outdoors, for my taste, lilies cannot be too sweet, and I especially love the strong Oriental scent of the auratums.

Fragrance, whether strong or delicate, is a highly subjective matter, and one gardener's perfume is another gardener's stink. My tastes are catholic. I very much like the pungent late-summer flowers—the marigolds, calendulas, and chrysanthemums, even the old-fashioned single nasturtiums that have not been prettied up by the hybridizers. These ranker autumn flowers, some of whose pungency comes from the foliage, are what Louise Beebe Wilder, in her book *The Fragrant Path* (1932), calls "nose twisters;" the very word "nasturtium" *means* "nose twister" in Latin. It is my habit to keep two little vases filled with small flowers on our living-room mantelpiece all summer; by September, these are often filled with nose twisters—French marigolds, miniature Persian Carpet zinnias, calendulas, and a few short sprays from the tall heleniums, all in the tawny and gold shades of autumn. But to some people the aromatic scents of these flowers and their leaves are unbearable. In fact, one friend of mine cannot tolerate them at all, and her pretty nose will wrinkle with disgust when she is in the room with them. If I know she is coming to the house in September, I hastily change my bouquets to the softer tones and sweeter scents of the late-blooming verbenas, annual phlox, and petunias.

I know next to nothing about fragrance. A year of trying to learn about it has left me as ignorant as ever, beyond a few simple facts that everybody knows, such as that a moist, warm day with a touch of sun will bring out fragrance, that hot sun and drought can destroy it, that frost sometimes releases it, and that rain will draw out the good chlorophyll scents of grass and foliage. The commonest complaint one hears today from the amateur gardener is that modern flowers, particularly roses, are losing their fragrance, thanks to the hybridizers' emphasis on form and color, and it had seemed to me, too, that many flowers smell less sweet than they used to. Sweet peas: The ones I grow are sweet, of course, but I remember them as sweeter still in my aunts' flower garden of long ago. Iris: It is rare to find a fragrant flower among the hybrids we now grow on our terrace, but a gardening friend with a particularly alert nose remembers the old-fashioned "flag lilies" of his childhood as very fragrant. Lilacs: The common single farmyard lilac for me has the headiest spring scent of all, and some of the double modern French hybrids are said to be as fragrant, yet the double lilacs we grow here are distinctly less sweet than our common lilacs (*Syringa vulgaris*), probably because one of their ancestors was the only wild lilac that has no scent at all. Pansies: Oh, the list of lost sweet smells could go on and on.

These vexing doubts have given me courage to write around, asking questions of the scientists and horticulturists and commercial growers whom I dared to bother with my inquiries. A few of them have answered helpfully, a few evasively, but most of them mysteriously. The only consensus I can gather is that fragrance is not now a strong incentive among the growers and breeders. Color and form and hardiness are the thing, because that is what the public demands. This may be so, but it seems to me that the least the seedsmen and nurserymen can do for those of us who value fragrance is to tell us in their catalogues more often than they now do when a flower is particularly sweet-scented. Take the case of the large white single hybrid petunia called Snowcap, which my friend with the alert nose notes as beautifully fragrant in his garden, despite the fact that it is a modern grandiflora. You may see Snowcap's handsome portrait in color in Vaughan's 1960 catalogue, and it is listed in "Park's Flower Book" for 1960 and described there as the "largest pure white." Neither cataloguer appears to think its scent worth mentioning, even though fragrance is rare in the large modern petunias. The Joseph Harris Company now sells Snowcap only to wholesale customers, probably because this year the firm has two even more beautiful large single white petunias to offer its retail customers—Seafoam and Snowdrift. These two are F_1 (first-generation) hybrids, so it could be they have no perfume. But perhaps, like Snowcap, which is also F_1, they have—who knows? Certainly not the catalogue reader.

The fragrance of the rose is probably more mourned when it is missing than that of any other flower, but it apparently would be inaccurate to say that modern roses are less fragrant than old-fashioned roses. Bertram Park, an English rose authority, writes in his *Guide to Roses*:

> Nothing irritates me more than the question "Why have modern roses lost their fragrance?" . . . There are two or three hundred species of roses . . . from which our modern roses have descended; of these only a half a dozen or so are fragrant. In spite of this handicap, the hybridists have succeeded in bringing out fragrance in modern roses.

The truth seems to be that, not counting the species roses, there are quite a few old roses without scent or with very little scent, and we all know there are plenty of new ones. There is a difference, however, in the fragrance of the old and the new. Each type of ancient European rose had its distinctive and beloved scent—damask, gallica, musk, alba, centifolia, and so on—whereas the scent of modern roses has become mixed by an infusion of many strains. This may be one cause of the mutterings about today's hybrids. I suspect, though, that the chief basis for complaint about their loss of scent derives from the prevalence today of the floribundas, many of which are still far more floriferous than they are sweet-smelling. We grow them for other reasons than scent. Even so, each year the rose men develop a few new fragrant floribundas. As a catalogue reader, I suggest that home gardeners would complain less were the rose men only forthright, coming straight out in their descriptions of roses with "Unscented" or "Very faint fragrance," if this is the fact, and never failing to mention perfume when it is present. If none is spoken of in the description of a rose, I now tend to assume that it has none, but this may be unfair. For example, Sutter's Gold, a modern yellow, which I grew last summer and which has one of the most delightful perfumes I know, is described in several catalogues this year without a hint of this great asset. I am also wary when no scent is mentioned because of a ludicrous experience I had with white roses. For years, we had growing here that favorite of our parents, a Frau Karl Druschki, a white hybrid perpetual without any scent at all. Finally I tired of Frau Karl's lack of fragrance, so I did not grieve when a very cold winter killed the old lady off. To succeed her I chose a white hybrid tea that I saw highly touted in the Jackson & Perkins catalogue. No scent, at least to my nose. The next year, I bought another, which J. & P. said was even greater. Still no scent. Neither was advertised as fragrant, but the praise was so loud and the pictures were so handsome that I simply failed to notice the omission. This year I see that the firm presents yet another "great" new white rose—the White Queen—but this time I shall be more cautious. Because there is not a word in the catalogue about the Queen's fragrance or lack of it, I shall avoid her and order instead Neige Parfum, which the J. & P. cataloguer says has "a delicious lemon-verbena aroma." Another white hybrid tea I want to try is Conard-Pyle's Blanche Mallerin, a French rose that is "nicely perfumed," and I certainly must order from Will Tillotson, for my bed of old-fashioned shrub roses, a

Mme Hardy, said in the catalogue to be the "finest white damask" and said by a rose-growing friend to have a unique perfume—a mixture of sharp almond and violet. Unique indeed! Fragrance is next to impossible to describe, and comparisons help very little; perhaps this is why the cataloguers duck it so often. What is sharp almond to one may smell like doughnuts to another. (A favorite catalogue adjective for scent is "fruity." What fruit, I always wonder.) As Colette said of peonies, whose fragrance someone had compared to that of a rose, "The peony smells of peonies; that is to say, of cockchafers."

Because there has been some misunderstanding among amateur gardeners, myself included, about the workings of the rose industry, I have been gathering a few of the facts about how the growers operate. Not all roses—or other plants or shrubs, for that matter—are grown at the address given in the catalogue. The biggest Eastern rose growers, like Jackson & Perkins, in New York, and the Conard-Pyle Company, in Pennsylvania, raise thousands of roses at their home nurseries, but they also grow many other thousands on the West Coast or in Arizona or Texas, where conditions are better for the rose. They do this in case of a disastrously cold winter in the East. The big companies usually have display gardens, where the customers may see the roses growing and select the ones they like, but not all good nurseries have such gardens. Nor do all good nurseries even raise their own roses. Some may have their own plants and budwood grown under contract, away from their home base, by wholesale growers. It hardly matters, for the process is about the same, and nowadays rose plants can be kept under refrigeration after they are dug, and shipped without damage, however far they travel. I think what does matter to the customer, though, is whether a catalogue address is merely that of an office or a shipping house, or is of a nursery where he will be welcome and may select his roses from a sample display. If he doesn't know, he may travel a good many miles in vain. Every firm could well state these facts quite plainly in its catalogue. Many firms do.

Unlikely as it sounds, the fashionable new colors in roses this year appear to be orange, tan, and brown. Jackson & Perkins's own Rose of the Year, which is featured in a striking color photograph on the firm's 1960 catalogue cover, is Hawaii, a brilliant coral-orange hybrid tea, said to have a pronounced raspberry scent. Just as the cataloguer says, Hawaii is "exotic," "sultry," "exciting." I really like the looks of its glossy dark leaves and big, broad, loose-petalled blooms, but I simply cannot imagine where I could plant Hawaii. It would look hideous in our terrace rosebeds, against the plum and lavender and purple of the clematis vines and alongside the traditional red, pink, salmon, white, and pale-yellow roses. Two other new J. & P. roses are Tanya, "radiant orange," and a floribunda named Brownie, tan and gold. The Conard-Pyle catalogue shows three fragrant roses in the same spectrum: Lady Elgin, "rich apricot and orange blooms"; Fantan, the first *tan* rose, which some people see "as a Chinese coral and even as burnt orange"; Sunlight, "Chinese yellow lightly washed with mandarin red." All these autumnal shades would harmonize, but I think the roses would need a bed apart, off near the marigolds and tiger lilies, and I suspect I would enjoy them more in October than I would in June.

The most ubiquitous rose of the year is not orange; it has "creamy tones with a subtle kiss of pink at the edge of each flaring petal." The name is Garden Party, and it is the All-America Rose of the Year. Garden Party, a hybrid of Peace and Charlotte Armstrong, can be seen in almost every rose catalogue but is best presented in that of the firm that bred it—the Armstrong Nurseries, Ontario, California. Despite that rosy kiss, I find Garden Party a bit insipid.

Some will know the catalogue of the Conard-Pyle Company under its title of "Star Roses." Though the roses this firm grows are indeed stars, its catalogue is not particularly notable for its beauty or for the quality of its color printing. Nevertheless, I have a great respect for Conard-Pyle and its publications. These include two catalogues a year and a biannual magazine for its customers, called *Success with Roses.* West Grove, the nursery's home ground, is near Wilmington, and here the firm raises more than half the roses it offers the public—an unusually high percentage for a big Eastern grower. There is a formal rose garden for display, as well as an acre, devoted to a "living catalogue," where a hundred varieties are planted in short strips, so that visitors can walk about and see the roses growing before they make their selections. In addition, twenty-five acres of gently rolling farmland are planted in long, contoured rows of young roses, which in September, when they are in the height of their bloom, must be a sight to see. I have never been there, but thousands make the pilgrimage to view them soon after Labor Day, when the company presents its new roses for the year. These are not roses it has hybridized itself but ones the firm has chosen from many sources, most notably from roses developed by the late Francis Meilland, the great French hybridizer. Conard-Pyle is the representative and distributor of Meilland roses in this country. Francis Meilland, who originated Peace and many other famous and even lovelier roses, died young, less than two years ago, but the roses he developed will continue to be introduced by Conard-Pyle for some years to come, and his father, wife and son are continuing his work. Conard-Pyle selects the Meilland roses it likes best each year and then tests them for suitability to our climate before it introduces them here. "Star Roses" shows on its cover this year, not quite satisfactorily, a beautiful Francis Meilland clear-yellow grandiflora, Golden Girl. The gayest of the 1960 Meilland roses, though, are the two prize-winning floribundas—Sarabande, a scarlet twelve-petalled rose that opens wide to show its bright-gold stamens, and Fire King, another flaming-red floribunda with double blossoms. Neither is described as having fragrance, but there is little reason they should have, since anyone who grew them would do so for a mass effect of strong color. Meilland's big peach-pink hybrid tea, Confidence, another hybrid of Peace, *is* fragrant, and it has a confident grace. Along with these new roses, the Conard-Pyle catalogue lists many of the old standbys, such as the sweetly scented reds, Nocturne and Crimson Glory, and some of the interesting older European roses, like Mme Cochet-Cochet, Mme Henri Guillot, and Suzon Lotthe. The thing I like most about this rose book is its candor. Of the old favorite Talisman rose it dares to say, "The blooms are sometimes a bit floppy, but Talisman devotees don't seem to mind"; of the classic Charles Mallerin, "not as many [blooms] as you'd like but . . ." Even of the proud new Meilland red hybrid tea, Royal Velvet, it admits that the blossom is so big it sometimes bends on its stem when it is wet. Honesty of this sort gives one faith. I have seen three recent catalogues of Meilland-Richardier, the French distributor of Meilland roses. The excellence of their color printing should put American engravers to shame, but because the French firm does not send its catalogues or sell its roses to American customers, the best we gardeners can do is to watch for the Meilland roses as they appear each spring and fall in "Star Roses."

The catalogue of the Bosley Nursery, Mentor, Ohio, has

the handsomest rose cover of the year. It shows against a deep-red background the acid-green leaves, the pure-white blossoms, and one elongated bud of June Bride. This ever-blooming rose, the first white grandiflora, was bred by Roy Shepherd, one of our leading American rose hybridizers. Mr. Shepherd's hybrids are always introduced by the Bosley Nursery. His less recent rose, Courtship, a watermelon-pink descendant of Peace, which you may also see in the 1960 Bosley book, is to me more attractive than the popular Pink Peace, so I was glad to learn that Courtship has been chosen by the American Rose Society as one of the ten American-bred roses to represent this country at the International Floriade, in Rotterdam, this coming spring and summer. Visitors are welcome at the Bosley Nursery, which is very much a family enterprise; the catalogue keeps one up to date on Bosley family news as well as on roses.

Judging by the catalogue addresses that follow, I can only believe that Mentor, on the fertile shores of Lake Erie, is our rose capital. I sometimes wonder how well these rival rose men get along as neighbors but have decided merely to be thankful for them all. So onward. Melvin E. Wyant (Johnny Cake Ridge, Mentor, Ohio) offers a good general selection of the better-known roses. The unusual feature of Wyant roses is that they are all grown in the North and are all *three-year-old* plants. Since most nurseries sell only two-year-old plants, I pass the word along, although I have never grown a Wyant rose.

The fat catalogue of Wayside Gardens (Mentor, Ohio) contains some of the best color photography in the American plant books, and this year's thirty-one pages of roses bring a rose garden into the house. The artist who takes most of the color photographs is Willard Kalina. Mr. J. J. Grullemans, the president of the firm, is, with his son, responsible for the readable and informative text. Customers are welcome at the nursery. I have lately heard complaints that Wayside has become too big and that its stock suffers from quantity handling. I have not found this to be true of its perennials, for which the nursery is most famed, since it carries a good many varieties not offered elsewhere. Its roses, too, have arrived in good order, perfectly packed. There are two Wayside hybrid teas this year that I like very much—Josephine Bruce, a black-crimson rose, which came from England, and Michele Meilland, a delicate pale pink with reddish stems, from France. (This was one Meilland rose the Conard-Pyle Company did not choose to introduce, but it was Francis Meilland's favorite rose and he named it for his wife.) I do not know whether either of these roses is fragrant; their scent or lack of it is not mentioned here or in the catalogue of the Bosley Nursery, which also lists these two varieties. To me, the loveliest pages of the Wayside rose panorama are those showing the small-flowered varieties. I'd like to grow the fragrant salmon-pink rambler named Albertine and a single-flowered pillar rose called Sparrieshoop, which is described as "mother-of-pearl pink." Of the floribundas, my choices would be Oberon, a soft salmon-apricot double, and Dusky Maiden, a dark, maroon single. Wayside also lists a few of the ancient roses, such as as gallica versicolor—the Rosa Mundi.

For a wider choice of the old-fashioned roses and shrub roses, there is the catalogue of the Joseph J. Kern Rose Nursery, Mentor, Ohio. Here, without fanfare and with the briefest of descriptions, are six pages listing a selection of nearly all the historical types. There is a particularly good list of the species shrubs and their more modern hybrids. For the convenience of the customer, every scented rose in the catalogue is sensibly marked with a capital "F," for fragrance. A prettier catalogue, giving an even wider selection of old-fashioned roses, is that of Will Tillotson's Roses, a California firm of which Dorothy C. Stemler is now the proprietor and genius. This pleasing book is Mrs. Stemler's own child, for she takes all the rose photographs and writes the text, and this year the cover is a color reproduction of one of her arrangements of old-fashioned roses in a tall vase. Her list includes the lovely Irish Fireflame, a single rose I discovered in a flower store more than forty years ago and have never happened to encounter since, in a store or in a catalogue. The two beds of Tillotson roses I planted two years ago did well last summer and should this year come into their own as shrubs. However, if anyone plans to make a similar planting, my advice would be to read first—which I failed to do—*The Old Shrub Roses*, by Graham Stuart Thomas, an English book (published in this country by Charles T. Branford, Boston). Now that I have read it, I may want to take care of some obvious omissions, like Mme Hardy, the white damask I spoke of earlier, and Camaieux, the most remarkable of the striped gallicas. Mr. Thomas is an authority on old roses, and has in his care the largest collection of them in the world. The chapters in his book on the evolution of the rose are by Dr. C. C. Hurst, the British botanist, and they give the clearest history of the rose I have read, tracing its progress from the wild species, through the European roses of the Middle Ages, to the beginning of the modern era, which started in the last decade of the eighteenth century, when the first four China stud roses were imported to the Western world, bringing with them the genes that make modern roses everblooming. (Unhappily, the genes that account for fragrance have not yet been isolated.) If you do read the Thomas book, you will find it a fascinating detective story, and you must be sure not to skip the section on the Apothecary's Rose, or Rose of Provins. It was in the little French town of Provins, in the thirteenth century, that the local apothecaries discovered a red rose that could be reduced to powder and still retain its fragrance. It is easy to guess where *that* led.

The best portraits ever painted of the old French roses

about which Graham Thomas writes are by Pierre-Joseph Redouté, the great botanical artist of the Napoleonic era. Redouté was born in Belgium; the Empress Josephine was his patron, and he worked at Malmaison for years, painting the roses in her garden. His rose folios were first printed between 1817 and 1824—one hundred and seventy plates. The Ariel Press, in London, published, in 1954 and 1956, a pair of folio-size volumes that reproduce four dozen of the Redouté plates, under the titles of *Pierre-Joseph Redouté Roses 1* and 2, with the original French text. I have seen only Volume 2, which has an over-ecstatic and none too accurate English introduction by Eva Mannering. The Redouté plates are what matter; every one of them makes a modern rose look clumsy.

To return to Will Tillotson's Roses: The roses Mrs. Stemler collects and propagates are grown elsewhere, but she has her small trial garden two miles from the office, near the shipping area in Watsonville, California. The business is ninety per cent mail order, so there is no display garden.

For anyone who lives in a climate warm enough to grow tea roses, an interesting catalogue is that of the Thomasville Nurseries (Thomasville, Georgia). It has a long list of these sweet-scented old-fashioned varieties that do so well in the Deep South. Some of the hardier of them may live as long as six or seven years as far north as Washington before they are killed off by a severe winter. If anyone wants to experiment with them in the northern or border states, the ones to try, according to Mr. Sam C. Hjort, who heads the firm, are Baroness Henriette Snoy, Mme Lambard, and Duchesse de Brabant, all three of them pink. He also offers many old-fashioned climbers whose very names fill me with envy—the Cherokees, the Banksias, Maman Cochet, Marechal Niel. Visitors are welcome at Thomasville and may view not only the nursery but the adjoining All-America Rose Test Garden, which Mr. Hjort directs.

At the opposite extreme, for gardeners in the very coldest climes, there are the famous Sub-Zero Roses, grown by the firm of Brownell Roses (Little Compton, Rhode Island). These are a special hardy breed, said to live without protection at fifteen below zero and to survive at thirty to forty below if covered. They were developed by the late Dr. and Mrs. Walter D. Brownell, whose son carries on the nursery. Sub-Zeros come in hybrid teas, floribundas, climbers, creepers, fountain roses, and tree roses. Eight varieties are non-susceptible to black spot, and all are said to be remarkably immune to disease. Most of them are rather curly roses, a kind I do not much care for, but a few are very pretty even so, and every one is a notable horticultural achievement. Three of the names are Orange Ruffels, Yellow Ruffels, and Lafter, so Dr. Godfrey Dewey, with his simplified spelling, seems to have been in there pitching on the sub-zero line.

The rose has been nominated as the national flower by popular vote. It is now up to the Congress to decide. If the choice is made official, I hope a native wild rose will be stipulated—say, Rosa virginiana, the common pasture rose, which flourishes from Maine to Florida and Texas, casting its spicy fragrance on the breeze along our lanes and highways.

Oddly, the most interesting seed books on my desk this winter are dated 1880, 1890, 1893, 1894, and 1897. All except the last are the work of Joseph Harris, and the precious collection was lent me by courtesy of his grandson, the present Joseph Harris, who is president of today's Harris Company, which is still situated at Moreton Farm. (In this country of change, and of scattered families and flight from the soil, the seed business shows a heartening continuity. Breck's of Boston, established in 1818, is still headed by a Breck; the sons of George W. Park run the seed business their father started in Greenwood, South Carolina, in 1868; David Burpee, president of our largest seed company, continues the plant breeding his father, W. Atlee Burpee, began in 1876; Vaughan's, Chicago's great seed house, also eighty-four years old, is still directed by men named Vaughan.)

The original Joseph Harris, who this year is my favorite catalogue writer, came to this country with his family from Moreton, England, in 1863. The travellers were on their way to New Zealand, where they planned to settle, but they stopped to visit friends near Rochester, and Joseph liked the look of the countryside so well that he bought a farm, named it Moreton, and stayed there the rest of his life. At first he grew vegetables for sale, but he was so good an agriculturist that as time went on his strains of seeds were in demand from neighboring growers, and gradually the seed business became more important than the farming. There was also in Joseph Harris another element—he was a born editor and writer. (The two interests often seem to go together—a talent for the soil, a taste for writing and editorializing.) The year Joseph Harris bought his American farm, he sold his first article to the *Genesee Farmer*. He tells about the sale of that article and about his subsequent life in publishing in the 1893 catalogue, under the heading "Forty-two Years' Work." They were forty-two crowded years. During them, he served as an editor for a year each on the *Rural New Yorker* and *Country Gentleman*. Then, for nine years, while he was a farmer and grower, he was also proprietor and editor of the *Genesee Farmer*. This work he gave up to become editor of the *American Agriculturist*, for which he wrote, until his death, in 1894, a column called "Walks and Talks on the Farm," which I am told is still remembered. He was appointed professor of agriculture at Cornell in 1867, but apparently could never tear himself away from the farm to fill the post, although the university hopefully continued his name on its roster for three years. Meanwhile, he was writing books. I know the titles of two of them, because they are advertised on the inside front cover of the 1880 catalogue—*Talks on Manures* and *The Pig: Rearing, Breeding, Management and Improvement.* That he could manage pigs and other stock is evidenced by advertisements, on the inside back cover, of "Moreton Farm Essex Pigs, the Largest Herd in the World!" and "Moreton's Pure Bred Cotswold Merino Sheep," which are illustrated by fine engraved portraits of an enormous porker and a sheep in full fleece. In the end, seed propagation won out, but Joseph didn't get around to publishing his first seed list until 1879. He distributed thirty thousand copies, yet he thought so little of it that no copy remains in the company's archives. The second catalogue, which I have before me, is a modest forty-eight-page booklet, five and a half by eight and a half inches, titled—in Old English lettering on a dove-gray cover—"Select List of Field, Garden and Flower Seeds For 1880. Cultivated and For Sale by Joseph Harris." There are no illustrations on this front cover, but the back cover carries an engraving, which is so quaintly pleasing one could well frame it, of Harris's Improved Yellow Globe Mangel Wurzel. All through the book there are delicate engravings of equal quality, including one showing three of the nine varieties of cabbage Joseph Harris offered that year. He did not merely offer them; he wrote a full page on how to raise them, discussing the merits of the several varieties and telling his customers how best to make a profit from each. The second cabbage on the list caught my eye:

EARLY JERSEY WAKEFIELD, the most popular and earliest market Cabbage; good size and sure to head. Per oz., 40 cents; per ½ oz., 25 cents; per packet, 10 cents.

I knew it well! Turn to Harris's 1960 catalogue and there it is still, illustrated by one of those clear photographs the modern Harris company does so well, above the caption "The tenderest, most delicious cabbage of all." The first Joseph Harris would not agree; *he* thought the Early York the best for home use, but said it was too small and too good for market. Such forthrightness must be a family characteristic, for today's Harris catalogues continue to make comparisons of varieties and to indicate the firm's favorites among the seeds it offers. In sixty years, the price of Early Jersey Wakefield seed has exactly doubled—a moderate advance, as things go.

A couple of other vegetables from the 1880 booklet have survived the years—the Hubbard squash, which the current Harris catalogue lists as the "true, original strain" of Hubbard winter squash, and the white summer squash, which appears to be the same today as it was then. In 1880, though, it was called the Early Bush Scallop, and now it is called the Mammoth Bush Scallop. While there are a couple of dwarfs in Joseph Harris's 1880 catalogue, there isn't a mammoth or a giant in the lot. The first giant to show up is the Giant Crookneck squash, in 1893. By 1897, there is a Mastodon carrot, but Joseph Harris did not invent the name; he had been dead for almost four years.

"I am not ashamed to own that I am very fond of flowers," he writes in 1880, "but I do not want a great many kinds. . . . Of these I want a liberal supply." He lists fifteen flowers, among them balsam, mignonettes, pinks, sweet peas, sweet alyssum, and the other small, fragrant annuals of the gardens of my childhood. Ten years later, he is listing three times as many, importing flower seed from Germany, urging farmers to beautify their grounds with flowers, offering seed at a discount to children and to clubs of buyers, and announcing the happy fact that Moreton Farm is now a post office. Joseph Harris obviously knew how to sell seeds as well as grow them, but if he was a clever merchandiser, he was also a thoughtful one. In this same catalogue of 1890 there is a vigorous paragraph headed "Better Times," and I quote it in full as a memorial to this versatile American seedsman and his era:

For the past ten years or more cultivators of the soil have had hard times. This is the case not only in our own favored country but throughout the world. This is not due to bad crops but to low prices. Farmers have had a terrible struggle for existence. There has been no overproduction. That is a mistaken notion. No farm produce has been wasted. It has all been consumed. But buyers, knowing that there was enough, held back and forced down the price. Consumers now know that this was as bad for them as for the producers. When farmers have no money they cannot buy. There has been enforced economy on the farm and extravagance in the city. We are, or soon shall be, the wealthiest nation on earth. We have no great standing armies to support. We are a nation of workers and such workers as the world has never seen before. We do not plow with a cow and a stick of wood. We no longer quote Franklin's maxim, "He that by the plow would thrive, himself must either hold or drive." We do both. We drive three horses abreast, and do more actual work in a day than was formerly done in a week. And in fact many farmers, not content with this, put on two teams of three horses each to a gang of plows, harrows, etc. Hitherto the consumers have had all the benefit of the producers' enterprise and industry. But the time has come when they will share it with us. The cities are rich and they are spending freely. They are not hoarding. The vast accumulations of money in our savings banks belong, not to the rich, but to the industrious classes. Many a young man and young woman put the money there till they can join hands and join purses and buy a small farm for themselves or start in business. Foreign labor has sought our shores for many years and now foreign capital is coming by the millions. All this means better times for farmers, and when farmers do well the country is prosperous. Let us be hopeful and go ahead with our improvements.

The first improvement Joseph suggests in the paragraph that follows is to enhance the beauty of the farm with lawns, gravel walks, evergreen screens, fruit and ornamental trees, and a few choice flowers, concluding with the remark that such surroundings improve the tastes and characters of the people and promote "industry, sobriety, enterprise, and every good word and good work."

Industry, enterprise, and every good work, including (as I have said) good camera-work and forceful text, characterize the catalogue of the Joseph Harris Company today. Most of the sparkling photographs are taken by Charles B. Wilson. The present Joseph Harris writes much of the catalogue, and his strains of seeds are as renowned among his customers for their quality as his grandfather's were in their time. The big news of the 1960 catalogue for me is that the color pages of flowers are for the first time as satisfying and true to life as the black-and-white photographs; I never saw better color in a seed catalogue. As for 1960 news of vegetables, there is a new butterhead lettuce, called Sweetheart (listed also by several other seed companies), which Harris says is better than its White Boston lettuce. I didn't think anything could be, but since Harris says so, I am forced from experience to believe it, just as I believe from experience that—as the catalogue says—Fordhook U.S. 242 is "the Best Bush Lima" and Tendercrop "the Best-Flavored Green Bean." Such superlatives are not boastful; they are merely helpful to a seed shopper.

Of all the catalogues of flower seeds, "Park's Flower Book" makes the best reading. You can study it endlessly, finding each time new surprises and new oddities you will want to try. The catalogue is rearranged this year, and the familiar old first page is missing, but far along in the book, on pages 64 and 65, Geo. B. Park, the president, writes his usual warm letter to his customers. One of the interesting introductions this year is a double gerbera called Fantasia; here is one case in which increasing the number of petals has not detracted from the charm of the flower. Park carries several varieties of single gerberas, too; after all, the firm's slogan is "All the Best of New and Old in All the Flower Kingdom." Last summer I grew a packet of mixed seed of some of Park's California Giant zinnias, and found them the best mixture I had ever had—fine, strong flowers and soft harmonizing colors, with all the jarring oranges, purples, and magentas screened out. I also grew Park's Sugar Plum petunia and was charmed. Do not judge it by its color picture in the 1960 catalogue, for something appears to have gone wrong with the color of the inks. Instead, turn back to the 1959 book, if you still have it, for a picture that shows this interesting petunia in its true rosy plum-pink coloring, or look at it in Harris's 1960 color pages. I also very much like Park's Blue Lace petunia, which I grew, and I shall have them both again.

The battle of the snapdragons is on this year. They are Burpee's featured flower, and three new F_1 hybrids named Burpee's Supreme are shown in full color on the handsome cover of the Burpee catalogue. Their individual names are Venus, Vanguard, and Highlife, and they are lovely in their soft-pink, rose, and cream-colored ruffled dresses picked out with notes of soft yellow. Vanguard, the rose, which grows three feet tall, won an All-America award. But hold! Snapdragons are also *Harris's* featured flower. Its two-and-a-half-to-three-foot Rocket snapdragons are likewise true F_1 hybrids, and there are *six* of them, and *all six* won All-America awards. Their "six velvety colors" are white, yellow, pink, orchid, red, and orange. Harris devotes a full page to Fred Statt, the company's flower breeder, who, in cooperation with Pan-American Seeds, developed these hybrids and who, as an individual, was given the unusual honor of a general award for outstanding achievement in horticulture by the All-America Selections Council. Harris does not desert its fine 1958 Panorama snapdragons, but it is content to offer only two kinds, plus one semi-dwarf variety. This is characteristic of Harris; it knows a good thing when it sees it, and enough is enough. Not so Burpee, which believes in quantity as well as quality. It offers the seed of Fred Statt's Rockets as well as of its own new Supremes, and a new double called Cream Puff, and mixed new doubles in pastel colors called Glamour Shades; these 1960 varieties occupy one page. Turn the page and there are two pages of Burpee's older snapdragons—inevitable, I suppose, since last year the big thing in snapdragons with Burpee was its Sentinel Snaps, and the year before that was the big year of its Tetra Snaps. ("Snaps" is Burpee's word, not mine. I detest the cozy flower abbreviations. "Mums" is probably the most repellent of the lot, unless it is "Glads," but 1959 gave us a new nasty—"Dels," for the lordly delphiniums.) Naturally, Burpee could not desert these recent strains of snapdragons so soon, nor should it have. Last year I grew Sentinels, and they were handsome. But this isn't all. There are Burpee's Tip Tops, four feet tall, and Burpee Giants, now mere pygmies by comparison, and its Extra Early Hybrids, and many, many more. And how does *Park* stand in the battle of the snapping dragons? Well, *it* features Rockets and the prize-winning Vanguard, and offers still another line of F_1 hybrids, called Hit Parade, which are shorter than Rockets but come two weeks earlier. As the catalogue says, the two "compliment" each other and should be grown together. Park is high, too, on the Harris Panorama strains, which it has renamed Ginger

Snaps. Vaughn's is on the sidelines of the fight, listing only the Tetras and some miniatures, but even so it all adds up to an embarrassment of snapdragons.

On my seed list this year, there will be, as always, an item for one large packet of small ornamental (ovifera) gourds, in mixed colors and shapes. I don't know anything else I can buy for so little money that will give me so much pleasure from early June, when I drop the seed into the ground, until the following March, when I usually decide to discard, as too autumnal, the gourds I have harvested and polished in early October. Right now, there is a bowl of last fall's crop in the center of the dining table, and a group of ten or a dozen more gourds decorates the living-room mantelpiece. I never seem to tire of the compositions made by the colors and odd forms of these cousins of the squash: the smooth egg- or pear-shaped white ones, the sleek dark greens, which can be solid color or striped in self-color or white, the yellows, the oranges, the creams, the parti-colors—smooth and warted—and all in an interesting variety of shapes that change a little from year to year. Because I have waxed and polished them myself by hand, instead of buying those too shiny varnished gourds the florists sell, they have a lovely soft sheen, like the patina of old polished wood. Growing gourds is easy if you know a few small tricks. Plant them in full sun, and plant more than you want, in order to have a good selection of color and form. Plant them well away from your squashes, pumpkins, and melons, or there will be cross-pollination and you'll find, as we once did, an inedible monstrosity growing in your vegetable garden. The vines will need watering in dry spells, and the gourds must be covered at night when the first frosts come. When we let them ramble on the ground, we just throw some grain sacks over them on cold nights. When we grow them on a trellis, I cover each gourd at sundown with a paper bag. (There is no more pleasingly comic sight than a paper-bag vine.) Gourds are ready to harvest when the stem that holds the fruit to the vine is dry and snaps off easily. In our climate this moment is apt to come just as the World Series starts. Thus, about October 3rd, having picked and then washed each gourd with diluted rubbing alcohol, to kill the molds and viruses, and having coated each one with floor wax—paste wax, not liquid—I settle down peacefully to buff the gourds with a soft rag while I watch the ball games. It is perhaps my most relaxed moment in the year, combining as it does a sense of harvest with two of my favorite sports. All the big seed catalogues offer the mixed gourd seed, and I have had good luck with the mixtures I bought from Harris and Burpee. This year Breck's has a pleasing color picture of its De Luxe Mixture of small ornamentals and also offers a special variety called Aladdin, which is turban-shaped with red and yellow stripes. Park, the indefatigable, lists not only mixed seed but *eighteen* varieties of the small ovifera gourds, including one called the Ornamental Pomegranate, or Queen's Pocket melon, which the catalogue says is deliciously scented. I really must have that Queen's pocket piece, and I trust that luck will bring me Willie McCovey and Willie Mays as well when I come to polish it. Possibly someone else would prefer to try one of Park's large gourds—those called lagenaria—say, the knobby Cave Man's Club, or the odd, swan-shaped Dolphin, or the Bombshell gourd, "a sensation when 1-in. Gourd-like fruit bursts forcibly."

I once saw a hydrangea vine that had pulled a chimney down—a climbing *Hydrangea petiolaris*. The great bittersweet vine on our woodshed might do as much if we did not cut it back each year, but if one is in command of the perennial vines they can be a garden pleasure the year round. Our bittersweet attracts birds to its red berries in autumn and early winter; last fall a yellow-breasted chat, usually a bird of the thickets and not of the farmyard, gorged itself there for several days. In summer our clematis vines are our favorites, especially the large-flowering hybrids, which we grow against a cedar windbreak on our terrace as a background for roses, lilies, petunias, iris, and the big pots of fuchsias, lantanas, and begonias. The best clematis catalogue is that of James I. George & Son (Fairport, New York), the world's largest grower of this subtle and fascinating vine. I understand that the firm grows the clematis sold by many of the big nurseries, but you and I can also get our vines direct from Fairport, and I can testify that they arrive in top condition. The booklet is gay with color, most of it good, but not every picture is entirely true to the live clematis, especially in the shades of blue and lavender. A note on the inside of the front cover acknowledges as much; I like that. Jackmani, the big four-petalled purple, is the best known of the large-flowered varieties and one of the easiest to grow, but last summer I fell in love with one named Mme Baron Veillard, whose six-petalled flowers are of a most subtle plum-rose color. The catalogue calls it lilac-rose, but I think it is more plum than lilac. Anyway, the Veillard color blends well with Jackmani (on its right) and with Mrs. Cholmondeley, "a rare lavender-wisteria blue" (on its left). Mrs. C.'s picture, by the way, does not do justice to the vine; with us, anyway, the flowers were a far lighter, clearer blue. One of the charms for me of clematis blossoms is that even while they are fading they are pretty, and after the petals drop the stamens continue on the vine as little fluffs or stars of white, yellow, brown, or black. The George listing helpfully designates the months in which each variety blossoms, so that one may plant for continuous bloom from June to October. A small-flowered clematis I like very much, which we grow along the fences and on the kitchen porch, is Mrs. Robert

Brydon, which is a pale, watery blue, making a very pretty effect when the blossoms are out. This is a sturdy vine, presenting no problems at all. I bought ours from Wayside Gardens, and you can see Mrs. Brydon's color portrait in the Waysidc catalogue. (George carries it but does not show its picture.)

If it is another sort of vine you fancy, you will also find in the George catalogue honeysuckles, ivies, trumpet vines, wisterias, Dutchman's pipe, bittersweet, polygonum (that quick-growing feathery white vine that thrives in New York City), and the powerful but decorative climbing *Hydrangea petiolaris*. Wayside has fewer clematis to offer than George, but it lists all the other vines in greater variety, including the handsome pyracantha, or fire thorn, which is really a shrub that can be espaliered rather than a vine. I've always hoped to grow a fire thorn, because of its rich red winter fruits and evergreen leaves, but I got discouraged when the first we planted was winter-killed. This was *Pyracantha coccinea lalandi*. Now I shall try again, for Gardenside Nurseries has a new and hardier type, *Pyracantha Kasan*, which was developed by the Arnold Arboretum, and which *Gardenside Gossip*, the newsletter of the Vermont nursery, reports survived the killing winter of 1958–59. It is like this nursery to find an especially hardy variety. Most famous for its own strains of hybrid pink and white trumpet lilies, and for its wild-flower plants and seeds, Gardenside makes interesting finds in many fields and reports on them in its little sheet, *Gossip*, which is now published only once a year, in late August. (White Flower Farm, always alert, also lists *Pyracantha Kasan* this year.)

The catalogue of perennials that most took my fancy in 1959 is the charming forty-six-page book issued by Lamb Nurseries, Spokane, Washington. (Display garden; visitors welcome.) The 1960 cover is not quite as delightful a composition, perhaps, as that of 1959, which showed in delicate color four English bedding violas—Better Times, a creamy yellow; Enchantress, bright lavender; Mt. Spokane, pure white; and Eileen, a soft royal blue—but you may see these lovely tufted pansies on the inside back cover of the 1960 edition. Lamb, like Gardenside Nurseries, sells my favorite viola, the blue Maggie Mott. Except for four pages of color on the inside and outside of the front and back covers, the Lamb catalogue is illustrated by excellent monochrome photographs. Attractive as the book is, the best thing about it is that it offers plants not available elsewhere, many of them real rarities—for instance, androsace, the choice rock jasmine, and onosma (Golden Drops), for the rockery; venosa, or *hardy* verbena, in three colors; *Aconitum anthora*, a *yellow* monkshood; *Phygelius capensis*, the Cape fuchsia, which is seldom seen; glaucium, the horned poppy; and the rare pimelia (New Zealand daphne). The roll call of the plants that *I*, at least, did not know could go on and on. There is, too, a very long list of bearded iris, which includes Double Eagle, a coppery-chartreuse double iris—the first *double* bearded iris, says the catalogue text. You may see Double Eagle on the 1960 cover, and it is lovely, perhaps because only the upright petals, or standards, are double. The chrysanthemum list is equally rich, and it includes spoons and rayonantes. I was particularly pleased to find eighteen varieties of true violets, some of them rare, such as vilmoriniana, a fragrant apricot violet from Europe, and Alice Witter, white with a red eye. "The popular belief that Violets are shade loving is a fallacy . . . with few exceptions," writes the cataloguer, adding that all eighteen kinds listed by Lamb prefer the sun. Someone who knows what he is about seems to be in charge of this nursery and this catalogue.

White Flower Farm's "Plant Book" came out late this

year, so I can only give it short shrift, but I find it as handsome and thorough a dictionary of the nursery's wares as ever. This year there is a touch of color on the cover and a lot of charming new line drawings by Nils Hogner. Last summer I grew several of White Flower's rare fuchsias, which were shown in color in the November-December, 1959, issue of *Notes*. My favorite was Whitemost, pale pink and white, without a hint of purple, but the others were lovely, too. For hanging baskets, Swingtime and Mrs. Victor Reiter are the best varieties—no fuchsia purple in them, either. Good news for pot gardeners is the new small sizes of Terra Rossa ring pots, which are made by hand in Italy for White Flower Farm. Visitors are welcome at the Litchfield nursery, and there is a two-acre landscaped garden. Although much of White Flower's stock is grown elsewhere, there are eighteen acres of shrubs and perennials under cultivation at the home nursery. I have had good luck with plants from White Flower Farm and particularly value its continued search for the unusual varieties in every line. This year it offers a new oddity that gourmets may want to know about—Buffalo Grass, from Russia, which is used as a flavoring for vodka drinks.

The dahlia is a flower I have never been able to make up my mind about. My father grew it, and in his old age was interested in the luxurious new varieties, so, for sentiment's sake, I have sometimes felt I should like it more than I do. I have often grown and enjoyed the old-fashioned pompons, whose neat and intricate quills appeal to my sense of order, such as it is. A couple of summers ago I decided I ought to be more experimental and try some of the big moderns, so I bought a half-dozen roots and planted them in the vegetable garden. Only two came up, but with just the two the results were embarrassing, because the flowers were so enormous. One was bright red and the other bronze, and both were as big as dinner plates. The only word for them was vulgar. I gave the roots away. Just the same, I have to grant that modern dahlias, not in the enormous sizes and if planted in groups, can in a short time make a very decorative show. The gayest American dahlia catalogue I have seen this year comes from Swan Island Dahlias, Canby, Oregon. It is a thin booklet illustrated mostly in color, and scattered through it are good hints on growing dahlias. The double page of cactus dahlias is full of lovely blooms in particularly subtle shades. But watch out! If you don't want dahlias of dinner-plate size you must buy the B sizes, which produce flowers no larger than four to eight inches. The A-size blossoms run eight inches and over, mostly over. The dahlia catalogue I really like, though, comes from Fa. D. Bruidegom, in Baarn, Holland. It has no color pictures, but the black-and-white photographs are so clear and pleasing that after reading the descriptions you can almost see the colors. (The text is in English.) The book, well printed on glossy paper, is fifty-six pages long and lists hundreds of varieties of every type of dahlia, with accurate descriptions. You are told how tall the plant grows, and a complex series of letters indicates whether the blossom belongs to group A (large), B (medium), BB (small-medium), or M (small), and whether it is, by the American classification, FD (formal decorative), ID (informal decorative), SC (semi-cactus), St C (straight-cactus), IC (incurved cactus), Pom (pompon), Ba (ball), M Ba (miniature ball), or An (anemone). I hate to tell you that a *second* set of letters gives the British classification, which is totally different, involving such combinations as Dw B (dwarf bedder) and SandF (double show and fancy). But the Dutch are orderly, and if you can once get in your mind which set of letters comes first, you can figure your way out of the maze. Prices are given in dollars, pounds sterling, and guilders. To import dahlias into the United

States a permit is necessary from the Department of Agriculture for every package of roots, but the company makes it easy by including an application blank to send to the Plant Quarantine Branch, in Hoboken, which in turn will send you an import tag to forward to Bruidegom with your order.

If sending to Holland sounds too difficult, you may prefer to travel by subway to the Parrella Dahlia Gardens, the Bronx. As I write, its 1960 catalogue is not yet ready, but I have just received a new price list, together with a copy of the 1953 catalogue. In both, the firm's proprietor, Mr. Albert Parrella, urges customers to visit his gardens when the dahlias are in bloom. He sells roots, plants, seed, and cut flowers in season, and his dahlias have won many awards. The old catalogue, with its dim photographs, is not handsome, but I note it because it is the first one I have ever received from a nursery within the city limits that is open to visitors. Then, too, I was pleased with some of the names and prices of Mr. Parrella's 1953 beauties: Mrs. Dean Acheson, $3.50; Lynn Fontanne, $2.50; Alfred Lunt, $1.50; Postmaster Albert Goldman, $4; Ogden Reid, $1.50; Night Editor, $7.50. (Mr. Reid would surely have approved of this last differential.) Mrs. Acheson and the Lunts have disappeared from the 1960 list, but the last three names survive, at reduced prices.

If I am lukewarm about the dahlia, I am red hot about the bearded iris. I like it without qualification, and would not be without it in the garden. I happen not to care for the pink varieties that have become so popular lately, especially the horrid flamingo pinks, and prefer the whites, blues, violets, purples, yellows, and coppers, not only because they seem more like the flower in its natural state but also because they add such strong notes of color to the spring borders. I am not at all averse, though, to the ever-changing combinations produced by the hybridists. In fact, the infinite variations of color in the falls, the standards, and the beards of the new varieties are a part of their charm. Almost every color-illustrated iris catalogue that specializes in the new hybrids is a handsome one, since the bearded iris is the most photogenic of flowers. I shall mention only one, which is perhaps my favorite this year—that of Cooley's Gardens, in Silverton, Oregon. The gleaming illustrations show not only solo portraits of the beauties but also iris growing in masses or in groups in gardens and arranged in bowls. The iris hybridists whose originations Cooley's Gardens introduces are Hall, Kleinsorge, Gibson, and Riddle. One rhizome of one of their new varieties may cost as much as $25, but such high-priced ones are not a bit more beautiful than the ones that, because they are old favorites and in less short supply, cost only a dollar or two. The new Silvertone, at $20, a silvery pale blue with a tangerine beard, is exquisite, but so is Temple Bells, an apricot yellow with a heavy orange-red beard, at $2. If I were to choose one de-luxe Cooley iris, it might be Frost and Flame, a Hall hybrid that is pure white with a tangerine beard, priced at $12—a delicious flower. And if you are searching for unusual iris, don't forget the Lamb catalogue, which even offers *twice-flowering* iris, that bloom both spring and fall. Except fragrance, that second blooming is the only thing the modern iris up till now has lacked—in my book, anyway.

No-One-Knows-What Addendum: The Ransom Seed Company, of Arcadia and San Gabriel, California ("Either address prompt service"), is offering "Atom Blasted Seeds" to home gardeners and schoolchildren. Under a headline that includes the words "Atom Blasted Seeds Are Harmless," its 1960 catalogue explains the experiments in irradiating seeds that have led to better barley in Sweden and to Brookhaven's famous Long Island carnations, which bear red and white flowers on the same stalk. It goes on to say that Ransom is offering irradiated seeds to the public in order that the progress in new plant development can be faster: "Save seed from any funny-looking plants or cripples; from these one can expect greater changes." For school projects, the company recommends that "pupils be of High School age, old enough to understand irradiated seeds *are not to be played with or put in their mouths*." (The italics are mine.) One can buy, at $1 a packet, the blasted seed of flowers, vegetables, field crops, fruits, and trees. On such tree seeds the catalogue's comment is "No one knows, not the most informed scientist, what will happen. It is a mystery. One could get strange shapes, giants, dwarfs, different foliage, improved timber varieties. No one knows what." This cheery booklet costs ten cents and lists, too, a little of everything else, from non-irradiated seed to flying squirrels advertised as not likely to die of fright—even though the rest of us do.

5

War in the Borders, Peace in the Shrubbery

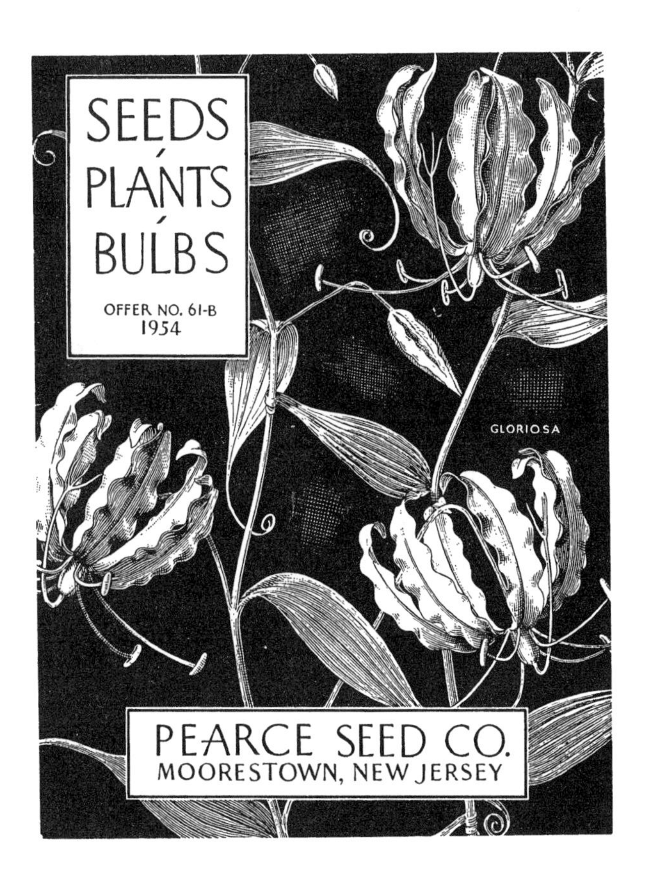

September 24, 1960

By August a flower garden, at least on the coast of eastern Maine, where I live, can be at its best—and at its worst. Most of one's successes are apparent, and all of one's failures. For me, this year, heavy memories remain from spring of the disaster area in the north bed of old-fashioned roses, where field mice, hungry under a snowdrift, stripped the bark off the bushes and killed two-thirds of them. Like all disaster areas, this one is still, although replanted, rather bleak. A more recent sorrow is the sudden death on the terrace of a well-established Jackmani clematis, which turned black overnight just as its big purple blossoms were opening. There are numerous theories in the household about this loss—too heavy a dose of fertilizer, too much watering, too strong a spray drifting over from the nearby rose beds, a disease still undiagnosed. My own theory is dachshund trouble. Our dachshund is a robin-and-bee hound, not a badger hound. A robust dog, he flings himself with abandon at birds, bees, and fireflies. Once he caught a barn swallow on the wing. Bees swarm all over the clematis bed, attracted by the petunias and violas, the foxgloves and lilies that we grow in front of the clematis, to give the vines the recommended "cool root run." I think the dachshund mortally wounded the Jackmani vine in a scuffle with a bee, for the other large-flowered clematis vines in that bed are spreading their mauve and mulberry stars all over the cedar windbreak, and the roses are in their second surge of bloom. The terrace, despite its accident, is one of our successes, and so, it would seem, are the long borders of perennials, with their masses of hardy phlox, which, because it is mid-August, are in full color. Yet a closer look at the borders will show that even here all is not well. The wars of aggression that I thought our private

Security Council and its little army of two, armed with spade, fork, and trowel, had settled in early spring have started again. The lolloping day lilies have begun to blot out the delicate columbines, the clumps of feathery white achilleas are strangling the far more precious delphiniums, and the phlox itself is at the throats of the lupines and the Canterbury bells. Even the low plants at the front of the borders are making aggressive sorties. The ajuga, whose small blue spires were so beautiful in June and early July, is one of the worst offenders. Unless I soon repress its insinuating roots, there will be no violets, pansies, or pinks next year. (I should have known better than to plant the stuff in the first place; after all, the ajuga's familiar name, bugleweed, carries its own warning.) There is also internecine warfare among the phlox—between the burgeoning clumps of common pink, white, and calico phlox and the less well-established stands of the newer varieties, whose colors are more interesting. Nonetheless, I am happy with all this bountiful bloom, and, careless gardener that I am, I comfort myself with the thought that at least I have achieved a mass effect, and that the flowers grow in drifts of color in a way that even Gertrude Jekyll, the author of *Colour in the Flower Garden*, might have approved. But that formidable garden genius of the last generation would never, never have condoned my crowded beds or my state of August sloth, which makes me want to say, "Oh, let it go. Let the plants fight their own battles."

It is in moods like this that a garden of flowering shrubs seems wonderfully easy and peaceful. Shrubs grow slowly. They need less care, less adjudication, less ruthless cutting back than perennials. We have never grown many shrubs here, probably because a well-landscaped shrubbery does not seem to suit our rural countryside. We do have a few, but they are the common ones, seen on almost every farm—lilacs, spiraea, honeysuckle bush, and shrub roses. Yet flowering shrubs are dear to me. I grew up in a house where the beauty of the shrubbery far surpassed that of the flower beds. We actually lived on a Hawthorn Road, in a suburb of Boston, and the street was named for the three huge English hawthorn trees that grew in our own yard—a red, a pink, and a white. Towering above the lilacs in the curving bed of shrubs, they were a sight to see in May. Only the most ambitious nursery catalogues seem to list hawthorn any more, and when they do they are apt to spell it "hawthorne," such is the carrying power of Nathaniel. To reach our May blossoms, we children had to carry a stepladder into the empty lot next door, use it for the first boost up, and then scramble the rest of the way to a narrow ledge on top of an enormously high lattice fence that backed our shrubbery. Standing there perilously, trying to keep our balance, we had to reach *up* to break the branches. Memory makes the fence at least twenty feet high, and the hawthorn trees many feet higher, but remembered Boston snowdrifts still tower way over my head, so perhaps our hawthorns were only the average height of eighteen or twenty feet. At that, they were the tallest of the shrubs.

Most people do not pick their flowering shrubs, but we always did. I can remember the succession of flowering branches, plucked by the adults of the household and arranged by them in a tall gray Chinese jar, in our gold-and-green parlor. My sister and I and our friends had a game we played with the shrubbery. It was called Millinery. All the little girls in the neighborhood would bring to our lawn their broad-brimmed straw school hats, which, because they were Boston girls' hats, had only plain ribbon bands for decoration. Then each of us would trim her straw with blossoms from the shrubs. There was a wide choice of trimmings—forsythia, Japanese crab, Japanese quince, mock orange, flowering almond, lilac, hawthorn, bridal wreath, weigela, deutzia, with its tiny white bells, and, in June, altheas and shrub roses. We were not allowed to pick the rhododendrons or the azaleas, but nothing else was forbidden. When our flowery concoctions were completed, we put them on our heads and proudly paraded into the house to show them off to our elders; it seems to me now that we must have made quite a gay sight. By dusk the trimmings were dead, and the next day we could start all over again.

Autumn, in most regions, is the best season for planting trees and all but the tenderest flowering shrubs, and I think that if I lived in a Boston suburb today—or anywhere in the East, for that matter—and were starting a shrubbery, I would send for the catalogue of the Weston Nurseries, of Hopkinton, Massachusetts, or, better still, would drive to the nurseries, on Route 135, seven miles from Framingham. Here there are several hundred acres of growing stock, plus a landscaped area with gardens and waterfalls and specimen plants of all kinds, where customers are welcome. The descriptions are sparse in the Weston Nurseries catalogue, and its eighty-six well-printed pages are illustrated by only a handful of decorative line drawings, but it lists an enormous number of things—evergreens, broad-leaved evergreens, trees, ornamental shrubs, roses, fruit trees, perennials, and much more. As an example of its completeness, I'll note that the 1960 list includes six varieties of hawthorn (spelled correctly), fourteen varieties of dogwood, fifty-three azaleas, eleven native rhododendrons, and twenty-nine rhododendron hybrids. I had two thornless honey locust trees come by truck from Hopkinton to our farm in Maine two years ago, and today they are thriving on our lawn, thus far unmolested by porcupines, which are the plague of the native locusts hereabouts.

Another excellent Massachusetts source of shrubs, trees, and evergreens is the Bay State Nurseries. Its current catalogue is unillustrated, and it lists fewer varieties than the Weston book does, but it gives explicit descriptions of all its shrubs and trees. A second catalogue Bay State puts out is an attractive fifteen-page illustrated booklet titled "Seaside Planting," which is full of helpful advice for anyone who wants to grow anything near the ocean—on sand dunes or wind-swept slopes, or where salt spray may reach the garden or shrubbery. Both of the Bay State catalogues are undated, which is inconvenient, for what do prices mean these days unless they are current? The stock offered in "Seaside Planting" includes not only trees, shrubs, and plants indigenous to the New England coast, like bayberry, sweet fern, shadbush, bearberry, dusty miller, which is called "the high-tide plant," and the delightful beach plum whose fruit makes such good jelly, but also exotic trees and shrubs that have proved stalwart against hurricane winds and flood tides. Bay State says that the Japanese black pine and the Austrian pine are made to order for our eastern coastline, that the Russian olive is useful as a windbreak, and that the spreading cotoneaster, with its pink blossoms, is quite as hardy as the rugosa rose, which is traditionally hardy near the sea. The other roses it recommends for the shore are the memorial rose (*Rosa wichuraiana*), the old-fashioned pink rambler Dorothy Perkins, and the American Pillar rose. For years we had an American Pillar growing against a trellis on the south side of our house. It is a single, deep-crimson rose with a white center and gold stamens. I had never quite liked its crimson as a background for the late tulips and the annuals that followed them, so I foolishly transplanted the big bush. Now I have my regrets, for it did not survive the move, and no other Pillar rose or climber has done as well in our prevailing southwest breezes.

A different way to start a shrubbery would be to send for the catalogue of the Kelsey Nursery Service. This is not a nursery but an office; no stock is grown at this address, except a show garden of broad-leaved evergreens. The catalogue explains that Kelsey's method is different from that of a nursery, and the firm makes no pretense of growing its own stock. It is a service, employing by contract specialist growers and, for wild stock, regional collectors, from all of whom the trees or shrubs or plants are shipped direct to the customer. This last is the difference between Kelsey and the large mail-order nurseries that contract to have their shrubs grown for them. Usually such contract-grown material is shipped to the nursery, then reshipped to the customer. Kelsey eliminates this intermediate step. Its theory is that the customer receives fresher material this way, at somewhat lower prices. The Kelsey catalogue is sensibly planned. It contains a climate map, with weather zones numbered from I to VII, keyed to the same numbers in the listings, so that you will not order anything unsuitable for the region you live in. It offers a wide variety of material—evergreens, deciduous trees and shrubs, fruits, roses, vines, ground covers, perennials, and wild native shrubs. The catalogue is not particularly attractive in appearance, but I was attracted to it by a section that contains an interesting selection of dwarf trees and espaliered dwarf trees. These espaliers are hard to come by. Kelsey offers them only in apples and pears, in three shapes—fan, U form, and four-armed. The list of standard fruit trees is much longer, and includes some unusual fruits, like beach plums, persimmons, and medlars, and also, I'm glad to say, the old-fashioned Red Astrachan apple, the best by far of the summer apples, which most growers seem to have abandoned. I have not yet used Kelsey, but it has been serving customers for more than eighty years.

We ourselves have usually bought our dwarf and standard fruit trees from the firm of Kelly Brothers, in Dansville, New York, which this year celebrates its eightieth anniversary with a good-looking catalogue in full color. Heretofore the

Kelly catalogue has not been one to charm the eye, but this year its Kodachromes of fruits, berries, and grapes are so luscious and so true to the natural colors that they make the mouth water. Apples, peaches, pears, cherries, and plums come in dwarf as well as in standard size, but no espaliers are offered—only a bulletin to tell you how to espalier your own fruit, which is not too easy a job. After eight decades, members of the Kelly family still run the nursery. Let us hope the handsome catalogue pays off well enough to be repeated. Its listings include nut trees, shrubs, flowers, and much more, but the name Kelly spells fruit to most of us.

A nursery that specializes in rhododendrons and azaleas is Lindum Gardens, in Portland, Oregon. Visitors are welcome at the nursery and may select their plants from among thirty-five thousand grown there. A startling photograph on the cover of the catalogue shows rhododendrons in noble bloom against a backdrop of Mount Hood, and all through the glossy booklet there are excellent photographs of the flower clusters of individual varieties. Unhappily, the 1960–61 catalogue is printed in green ink, and is therefore less pleasing than the 1958–59 issue, which was done in black-and-white. Green ink or no green ink, though, people who have bought shrubs there tell me they are of the highest quality. One useful feature of the booklet is the symbols designating the relative hardiness of the varieties, each one being ranked from H–1 (hardy to twenty-five degrees below zero) to H–5 (hardy to fifteen degrees above zero). We have never grown and cultivated rhododendrons or azaleas here, thinking them too elegant for a Maine coast farm. Besides, we have our own display of wild azaleas in May, when the rhodora (*Rhododendron canadensis*) spreads its purple-pink bloom in drifts across the pasture, washing up in waves of color against the green of the surrounding spruce woods. In this region, rhodora is called lambkill, but when we were raising sheep it never killed a lamb.

Another Oregon specialty nursery, the Brownell Farms (Milwaukee, Oregon), grows only ilexes—that is to say, hollies. The firm, started by the late Senator George C. Brownell in 1910 to grow holly for the Christmas trade, has been developed by his son Ambrose Brownell into a nursery and arboretum for the collection and propagation of hollies from all over the world. Until I saw the Brownell catalogue, I hadn't known there were so many different varieties of this cheerful shrub and tree. The nursery sells some seventeen species and well over a hundred named varieties. Most of them are English hollies (*Ilex aquifolia*), but there are also the decorative golden and silver hollies (*aurea* and *argentea*), with variegated leaves, which may have gold or silver margins, or gold or silver centers. Brownell offers, as well, hollies from the Orient, from the Canary, Madeira, and Balearic Islands, and from our own continent. Many hollies are hardy and will winter well in the North, surviving at ten to twenty below zero provided a cold spell does not come in early winter. This last lets out my own state of Maine. Thus, I cannot say, with Southey:

> *But when the bare and wintry woods we see,*
> *What then so cheerful as the holly tree?*

Ambrose Brownell can and does say it, in his first and current nursery catalogue, which is a handsome booklet, full of pictures of holly trees in brilliant color and in black-and-white, and full, too, of scientific ilex information, some of which is presented in rather fancy form. Send for it anyway, if you would grow a holly tree, or go to the Brownell Farms, a short drive from Portland, and pick out your own specimens. You may also view what Mr. Brownell likes to call his ilexetum. The smaller bushes are sold in pots; the large trees come to you balled and burlapped. You must never fail to plant two of a kind, for all hollies are dioecious, which means that some trees are male, with staminate blossoms, and others female, with pistillate blossoms. Unless you grow one of each sex you will have no berries.

If we can't grow a holly here in Maine, at least we have hopes of bright berries on our new fire thorn, which we planted last spring against the gray-shingled south wall of the barn. It is a *Pyracantha Kasan*, the extra-hardy fire thorn, which we obtained from the Gardenside Nurseries. The husky little bush has spread and flourished in spite of the dry summer, even in competition with a patch of ornamental gourds I planted too near it.

The firm of Edward H. Scanlon & Associates (Olmsted Falls, Ohio) grows conventional trees, but its specialties are rare and unusual trees and what its catalogue describes as "tailored trees." The term is misleading, suggesting as it does trees pruned and tailored into odd outlines, when all it means is trees that are smaller at maturity than most, or that grow, without pruning, in various odd shapes—globe-headed, umbrella, weeping, pyramidal, columnar, and so on. Scanlon recommends the short trees for city streets or moderate-sized house plots, where telephone or power lines may interfere with the branches of trees taller than twenty-five feet, and it recommends those of unusual form for special effects in landscaping. The list of trees is varied and imaginative, it is profusely illustrated with photographs, and it includes many flowering trees not to be found elsewhere. The catalogue I have before me is for the wholesale trade; I trust that the 1961 retail catalogue, due sometime in September, will be just as inclusive and as interesting to read. Nine varieties of my favorite hawthorn are offered, one of them the Glastonbury

Thorn, which, according to the legend, first sprang from the staff of Joseph of Arimathea. In England, there is also a story that the tree will burst into bloom when visited by a member of the Royal Family. The one planted near the Washington Cathedral, in the District of Columbia, is said to have done the trick twice—for the Prince of Wales in 1919, and for the present Queen in 1951. . . . Well, let's see, now. What were the *months* of those visits?

The catalogue of the Strawberry Hill Nursery is a pretty little booklet, devoted entirely to lilies, lilacs, and hardy camellias. The nursery guarantees only that these camellias will live as far north as Philadelphia, but adds that plantings of this especially hardy strain have survived cold winters a hundred miles farther north. We can't grow camellias here in Maine, but we have ordered two lilacs from Strawberry Hill to replace a couple of played-out old common lilacs, and have chosen from the catalogue's fine list of French hybrids Mme Casimir Perier, a sturdy double white, and Mrs. Edward Harding, whose double claret-colored florets are stained with pink. In a climate less severe than ours lilacs can be planted in the autumn.

Many of the nurseries one thinks of as primarily sources of perennials, like Wayside Gardens and White Flower Farm and the Lamb Nurseries, also sell shrubs; they are particularly good sources of the unusual flowering shrub. It was Wayside that introduced, for instance, the dogwood with bright-coral branches that is so decorative against the snow—Cornus sibirica Coral Beauty—and White Flower has recently patented a handsome new potentilla called White Gold, which it imported from England; its pale-yellow flowers are as big as fifty-cent pieces. Lamb's list of shrubs is short but select, and includes the tall, creamy rock spiraea of our own West, *Holodiscus discolor*.

Before ordering any new shrubs, an American gardener

would be well advised to get hold of a copy of *Shrubs and Vines for American Gardens*, by Donald Wyman (Macmillan, 1958). Dr. Wyman is the chief horticulturist of Harvard's famous Arnold Arboretum, and his book is the most useful reference book I know on these subjects. Plants are first discussed in terms of their hardiness, their order of bloom, their ornamental fruits, and their autumn coloring, and then follows a long list of the shrubs and vines the author can recommend. From there, he goes on to an equally useful "Secondary Plant List"—the varieties he cannot endorse, with the reasons for his dissatisfaction.

If in planting any of the shrubs or trees I have mentioned, you and I strive to obtain a natural effect, to follow the contours of the land, or to study the region we live in so as to make our planting suit it, if we naturalize garden flowers in our woodlands or bring wild flowers into our gardens or strive to make our garden blend gradually into a forest or field, we probably owe our ideas, though all unconsciously, to two great English gardeners and garden writers of the past. They are William Robinson, the author of *The English Flower Garden*, which was published in 1883 and is still a garden classic, and his younger friend and follower Miss Gertrude Jekyll, the formidable lady I referred to earlier. It was these two, more than any others, who taught Victorian England, and eventually America, to make the garden, as Mr. Robinson put it, a reflection of "the beauty of the great garden of the world." When he wrote this, most English gardens, except cottage gardens, either imitated the elaborately formal gardens of France and Italy or were devoted to "bedding out"—that is, to planting flowers in scattered small beds to make artificial color patterns or geometrical designs. Miss Jekyll has been called "the beautifier of England," and, indeed, during her long life she almost singlehandedly gave the gardens and estates of England a wholly new countenance. Between the years 1899 and 1925, she wrote fourteen books, all now out of print. For some months now, I have been reading three of the best of them, *Wood and Garden*, *Home and Garden*, and *Colour in the Flower Garden*, and also her nephew Francis Jekyll's memoir of his aunt. It has been a pleasurable but rather exhausting experience living with Miss Jekyll at Munstead Wood, the home she built for herself in West Surrey in her fifties, watching her plan and build the house itself with her architect, Sir Edwin Lutyens, and helping her carve garden after garden out of the birch and Scotch-fir forests and the sandy heaths that surrounded her house. There was no labor too heavy for her, no detail too small for her attention, no stone she could leave unturned—or, at least, unplanted with rock plants. Soon the gardens so proliferated that I lost track of them, and now remember, offhand, only the border of spring bulbs followed by ferns; the copse of daffodils; the spring garden of other flowers; the rock-wall garden; the hidden garden; the main hardy border, which was two hundred feet long and fourteen feet wide; the garden of summer annuals; the October garden of Michaelmas daisies; the primrose garden (primroses were a great specialty of Miss Jekyll's); the kitchen garden; and everywhere—on walls, trellises, and arbors—clematis, roses, and honeysuckle vines. The very size of it all is a bit stupefying. One year, her nephew tells us, eleven thousand roses were gathered for the potpourri, and to the roses were added half again as many sweet-geranium leaves and lesser quantities of lavender, sweet verbena, orange peel, bay, rosemary, and, of course, orrisroot and spices. (You will find the recipe in *Home and Garden*.)

Miss Jekyll, as you can guess, was the daughter of a gentleman of means. She set out to be an artist, but failing eyesight made her turn from painting, silver *repoussé* work, and fine embroidery to flower gardening. She had the means to hire a head gardener and, one supposes, all the helpers he needed, but she was always a working gardener herself, and in the garden wore army boots and an apron with great pockets for her tools. There is a painting of these garden boots of Miss Jekyll's in the Tate Gallery, by the artist William Nicholson, who also did a fine portrait of Miss J. at the age of seventy-seven, sitting semi-profile in an armchair, her tiny, thick-lensed spectacles on her nose, a velvet ribbon in her white hair, her artist's pointed fingers, now worn and wrinkled, pressed fingertip to fingertip in readiness for the next task.

What the lady who wore those sturdy army boots accomplished at Munstead, first at her mother's house and later at her own, was before long so notable that she was in demand as a garden consultant and as a designer of gardens for others; at her death, her commissioned garden plans numbered over three hundred.

The next-to-last chapter of *Home and Garden* is titled "Things Worth Doing," and I quote from it:

> We should not let any mental slothfulness stand in the way of thinking and watching and comparing, so as to arrive at a just appreciation of the merits and uses of all our garden plants. . . .
>
> There is no spot of ground, however arid, bare, or ugly, that cannot be tamed into such a state as may give the impression of beauty and delight.

And this is how the book ends:

> Are the people happier who are content to drift comfortably down the stream of life, to take things easily, not to *want* to take pains? . . . I only know that to my own mind and conscience pure idleness seems to me to be akin to folly . . . and that in some

form or other I must obey the Divine command: "Work while ye have the light."

I don't know whether there has ever been a Gertrude Jekyll anthology. If not, I wish some publisher would get one up, for while her books have a period flavor, most of what she wrote is not really out of date. She did, though, write too easily and flowingly, and her chapters vary in interest. Thus, a selection of the best from all the books would be rewarding. If to some of us the life at Munstead sounds more like slavery than like godly work, we should remember that to Miss Jekyll her labors were an expression of a discriminating love for flowers and a consuming artistic passion. It is therefore pleasant to note, offered for the first time, in Wayside's 1960 autumn catalogue, a newly imported periwinkle called Miss Jekyll's White—a free-blooming, small-flowered variety of *Vinca minor alba*, which the cataloguer describes as "a select and refined ground cover for particular locations." I can't help thinking that whoever wrote that sentence has, like me, been living at Munstead Wood with the author of *Wood and Garden.*

Well, while we have the light we had best be on to wild flowers. Autumn is the season to plant them, so there is not a moment to lose. Possibly I should report that I had perfect success with the spring planting of three potted trailing-arbutus plants from the Putney Nursery, which took hold in a damp end of the north border, under light shade from a nearby oak. Because we watered them, they have even survived this summer's severe drought. Putney's plants don't die if you take care of them, and Putney's customers are happy ones, for this is the nursery that never fails to notify you of the impending arrival of your orders. (I have lost a good many delicate plants because I was not at home on the day they arrived in the mail.) Putney's catalogue is an unpretentious booklet peppered with engaging flower photographs, and its eight-page list of ferns and the smaller wild flowers is my favorite of all the wild-flower lists. There are many temptations here—wild flowers I don't seem to know, like Oconeebells and black cohosh, as well as the dear, familiar exquisites like bloodroot, hepatica, lady's-slippers, Dutchman's-breeches (the wild dicentra that is easy to grow on a shaded woody slope), and the cardinal flower, the scarlet lobelia that loves the borders of streams and brooks. Whenever a wild flower (such as the large blue lobelia) will adapt itself to growing in a garden, the catalogue tells you this happy fact. The Putney lists include much more than wild flowers, offering a good selection of perennials, herbs, Vermont-grown hardy shrubs and trees, fruit trees, roses, and so on. Visitors are welcome at the nursery, which is on Route 5, midway between Brattleboro and Bellows Falls.

Although the chief specialty of the Gardenside Nurseries, where I got my fire thorn, is lilies, the new fall edition of *Gardenside Gossip* contains not only a lot of good lily talk but also a selective list of rare native plants and a column on the cultivation of wild flowers. The varieties of wild flowers offered by Putney and Gardenside overlap to some extent, but each offers rarities not on the other's list. This issue of *Gossip* is particularly interesting on the subject of pyrethrum (the painted daisy)—a garden perennial one finds in few American catalogues. Gardenside lists nineteen varieties and says that, unlike chrysanthemums, to which they are cousins, painted daisies should be planted in the autumn.

A wild-flower fancier, especially if he lives in the Midwest, should send for the pleasing catalogue of Lounsberry Gardens, in Oakford, Illinois. Here the illustrations are mostly in color and the listings are confined to hardy wild flowers and ferns, garden perennials, rock plants, and gladiolus bulbs. It has always seemed to me wise to buy wild flowers as near home as possible, and this is the only reason I have not yet bought plants from Lounsberry Gardens.

Another nursery I have never used, but one that is highly recommended by those who have, is the Gardens of the Blue Ridge, in North Carolina. It is probably most famous for its hardy wild rhododendrons, azaleas, and mountain laurel, and the other wildings of the Blue Ridge Mountains. Most of these shrubs can safely be shipped to distant points, and they will thrive in other cool regions. The nursery has been in business for sixty-eight years, and its illustrated catalogue is exciting, with its long lists of hardy native ferns, trilliums, and bog plants, as well as its wild shrubs. What I covet most, though, from Blue Ridge Gardens is its Turk's-cap lily bulbs, for the *Lilium superbum* is one of the most delightful of our native lilies, with its tall candelabras of nodding orange-red flowers. These lilies grow so prolifically at the four-thousand-foot altitude of the nursery that it can offer the bulbs at very low prices, especially when they are bought in quantities for naturalizing. Turk's-caps are accommodating. They will grow almost anywhere—in the garden, in meadows, woodlands, and swamps, and on rough ground—and they need next to no care.

Though I am a devoted reader of the spring catalogue called "Garden Aristocrats," which is sent out by the Pearce Seed Company, and recognize it as an excellent source for the seeds, plants, and bulbs of wild flowers, and for unusual perennials and house plants, I do find it a bewildering catalogue to order from. This is chiefly because, instead of being a small bound booklet twenty-two pages long, it is a folder with eleven faces, printed on both sides of each face, which folds up like a road map and stretches out, when opened, to a length of three feet eight inches. What with all the folding and unfolding, I get lost and am never quite sure whether I'm under the "Pot Plants" heading or the "Garden High-Lights" heading, or just where to find the wild flowers, which are scattered hit or miss throughout. Then, too, most of the type is irritatingly small—especially the minuscule letters that Pearce places after the names of the wild-flower seeds, which are supposed to tell you where and when to sow each variety. I become so utterly confused after flipping about from fold to fold to find the particular sort of seed I want that by the time I discover my item I have lost the table showing what these tiny letters mean and have forgotten the differences between *x* ("Sow after soil is warm") and *k* ("Sow in spring while soil is still cool"), *kt* ("Sowing can be in open ground in either late autumn or early spring") and—a very fine distinction here, it seems to me—*yt* ("Sow only in late autumn to earliest spring in open ground"). Nevertheless, on days of unusual calm I have managed to place orders with Pearce successfully, as some interesting geraniums in pots on the terrace attest. Next month, if I have time to search them out again in all those wandering pages, I may buy a packet of Pearce's mixed "Wild Flowers for Shade" and a packet of its mixed wild gentian seed. Pearce brings on the disease I call catalogue daze in a particularly severe way. I mention it, more in sorrow than in anger, if only because I have started to make out at least four orders for Pearce that I didn't have the stamina to complete. (Pearce's fall catalogue, "Autumn Rarities in Seeds, Bulbs, and Plants," which has this moment arrived in the mail, is the same sort of road-map affair, but I am glad to report that typographically it is less confusing. You will, though, need the spring folder for wild flowers, as only a few are included among the fall rarities.)

There is an unusual book on wild flowers, *Flowers Native to the Deep South*, which was published in 1958. The author and illustrator is Caroline Dormon, and the publisher and only distributor is Claitor's Book Store, Baton Rouge, Louisiana. Although all this may sound far too regional and special for most readers—and it *is* special—I found the book delightful, and I believe that other Northerners will enjoy it. Miss Dormon's more than two hundred illustrations, in water color and line, are both charming and exact, her text is lively, and, of course, many of the flowers she lists are also native to the East, North, and West. Miss Dormon lives in her own stretch of forest in northern Louisiana, where for more than thirty years she has been collecting and growing trees and wild flowers. Good as her botany is, she would probably not claim to be a professional botanist, nor does she attempt to make this a complete reference book of the flowers of her region, but she is certainly a gifted collector and a close observer, and her comments on the many hundreds of wild flowers she does describe are illuminating.

September and October are the months for planting peony roots, and before planning his peony bed this year the gardener may want to read a new book, *Peonies, Outdoors and In*, by Arno and Irene Nehrling (Hearthside). Oddly enough, it is the first non-technical book on peonies in more than twenty-five years. Useful and good-looking as it is, I wish it were better. Perhaps because of its chapters on peonies in flower arrangements and garden design, it has a garden-clubbish, amateur air that does a disservice to the rest of the book, which clearly explains the various types of peonies (tree peonies, herbaceous peonies, hybrids, and the rest), gives excellent advice on their culture, and supplies useful lists of the modern varieties that have proved themselves. The handful of photographic illustrations in color are lovely, and there are line drawings and diagrams. As a lazy gardener, I have always been grateful to the decorative and easy-to-grow peony, which was named for the physician to the Greek gods. The myth goes that Paeon, a disciple of Aesculapius, was the first to use the plant as a medicine, and with it healed Pluto of a wound he received in the Trojan War. Aesculapius, like many a modern teacher, was jealous of his pupil's achievement, and plotted Paeon's death, but Pluto, hearing of the plot, and grateful to Paeon for saving his life, in turn saved Paeon's by transforming him into a peony. The flower could as suitably have had a Chinese name, for it was known and honored in China from the most ancient days, and had its place in Chinese poetry and art by the eighth century A.D. You will find more of the peony's curious history than the Nehrlings give in Mrs. Buckner Hollingsworth's *Flower Chronicles* (Rutgers, 1958), and she explains more fully the peony's place in the kitchen and the medicine cabinet over the years. If this fall you decide to replace your old peonies with newer varieties, you might try eating the roots you dig up. The English nobility enjoyed peony roots in medieval days, finding their acrid taste a pleasant accompaniment to roast pork. I think, though, that if I ever discard any of my peonies I shall make them into "Compound Piony Water," following the recipe given in the Nehrlings' book, which is taken from *The Complete Confectioner*, by Hannah Glasse, published in 1782. It's a promising compound, all right, calling for many spices and one gallon of rectified spirits of wine. One can readily understand why Piony Water is "good in all nervous disorders" and "efficacious in all swoonings, weakness of the heart, decayed spirits," and other heavy afflictions.

At the moment, though, I shall restore my decayed spirits by studying some of the new peony catalogues and ordering some of the lovely modern varieties. The firm of Gilbert H. Wild & Son (Sarcoxie, Missouri) has long made a specialty of peonies, as well as of iris and day lilies, and its seventy-fifth-anniversary catalogue devotes nineteen pages to peonies, with inviting color photographs of scores of varieties of the herbaceous sorts that Wild offers. It sells no tree peonies, which are more difficult to grow, but there are two pages of decorative Japanese herbaceous peonies and a short list of the new herbaceous hybrids. The descriptions indicate the height, the flowering date, and—wonder of wonders—the degree of fragrance of each peony. Wild is sometimes even courageous enough to say so when a peony has no scent. Perhaps its example will encourage the rose growers to be equally bold.

The great advantages of hybrid peonies, aside from their startling beauty, are that they lengthen the period of peony bloom, many of them blossoming as early as mid-May, and a few even earlier, and that they introduce many interesting new colors to the peony spectrum—brilliant pinks, yellows, and unusual reds. Professor A. P. Saunders was probably the most adventurous of the pioneer hybridizers. Since his death, in 1953, his daughter Silvia Saunders has carried on both his nursery and his hybridizing experiments. Her new 1960 catalogue makes the sad announcement that it may be the last to come from the A. P. Saunders Nursery, of Clinton, New York. It is therefore a "must" for all who care for these rare hybrids. Miss Saunders writes me, though, that this is by no means her last year of running the nursery; she plans to sell peony roots and seed for at least three or four more years. Her little booklet, as I have said before in these pages, is merely a descriptive listing of her treasures, but it contains seventy-odd named hybrid tree peonies and about a hundred and twenty herbaceous varieties—an astonishing number when you consider how many years it takes to create a good hybrid.

Surely the most cheerful and hopeful of the fall catalogues are those of the spring-bulb growers. A year ago, I described the delights of the catalogue put out by P. de Jager & Sons, a firm that raises its bulbs both in Holland and in this country. Its 1960 version, with its many color illustrations, is to me still the most alluring and the most inclusive of them all, and I can testify that in every category—tulips, narcissi and daffodils, hyacinths, freesias, crocuses, and many sorts of miniature bulbs—the de Jager bulbs I planted last October performed miraculously. This year, I plan to try some of the firm's historical Broken Tulips. The word "broken" means that the flowers are feathered or flaked with bright colors or, if the background is dark, are marbled with white. These tulips, the cataloguer explains, are the ones that were so fashionable and brought such exorbitant prices in Holland during the seventeenth-century tulip mania, and they are the ones the Dutch and Flemish artists liked to paint. (One type of Broken Tulip is called the Rembrandt.)

Last June, as an experiment and on the recommendation of a knowing gardener, I ordered some of my daffodils for this autumn's planting direct from Holland—from J. Heemskerk. The bulbs are supposed to come direct to my door, all duties and other shipping costs paid, and I shall need no import permit. It is now too late to send an order to Heemskerk for this fall's delivery, but for another year you might like to send for the Heemskerk catalogue. It is less complete and less sophisticated than de Jager's, and occasionally its English is shaky and its spelling odd, yet the booklet has a charm all its own, and most of the color pictures are good. Whereas tulips are the great de Jager specialty, here daffodils come first. I was particularly pleased by the cover picture of Silver Chimes, a white hybrid narcissus of the triandrous group with a primrose cup, which has clusters of six or more flowers on each stem. It is perhaps *not* too late to order amaryllis from Heemskerk, who offers not only bulbs for the normal midwinter season of growth but also specially prepared amaryllis, in both named and unnamed varieties, that are shipped in the middle of October and should, if all goes well, come to bloom by Christmas. Just plant them in a pot and they should shoot. I hastily add that I've never tried these specially prepared bulbs myself and cannot testify to their performance.

The Little England Daffodil Farm, in Bena, Gloucester County, Virginia, sells only daffodils and narcissi. The 1959 catalogue, which is the latest I have at hand, shows on the front and back covers pictures of the old plantation manor house and the growing fields of the Little England Farm. Otherwise the booklet is just a list, but what a rich and unusual one! The descriptions are full enough to enable you to visualize the flowers, and only the most rigorously selected varieties are offered. The 1959 price list gives an idea of the rare quality here, for while there are some bulbs offered at as low as thirty cents each (if one can call that low), others range freely up to twenty- and thirty-odd dollars, and even higher. Glendermott, at $85 per bulb, and Empress of Ireland, at $125, must have come down in price by now, but they rather took my breath away, even as new introductions in short supply a year ago. The growing conditions at Bena are similar to those in Holland and Ireland, the catalogue says, and, to judge by the variety names, many of the parent bulbs must have come from Ireland. The emphasis here is obviously more British and Irish than Dutch; one of the useful features of the catalogue is the Royal Horticultural Society's classification, with diagrams, of the nine types of daffodils, so that you can tell exactly what you are getting when you order a triandrous hybrid, a cyclamineus hybrid, a Tazetta, a Poeticus, and so on.

I should add that I was charmed by the Bouquet, or

multiflora, tulips I grew for the first time last year. I planted them as the border of a bed of taller single tulips and daffodils, and the effect of the row of little clumps of tulips, each clump growing from one bulb, was very gay and airy. I bought the bulbs from the George W. Park Seed Company. Park's new autumn catalogue devotes almost two pages to these comparatively new tulips, and there you may see Pink Lemonade, Lovely Lady, Mahogany Prince, and all the rest of the Bouquets in pleasing color and black-and-white photographs, as well as a good selection of other spring bulbs. The prices are moderate. I might also note, in passing, that Jackson & Perkins appears to be selling the favorite pink-cupped daffodil Mrs. R. O. Backhouse this autumn under the name of Mrs. R. O. Pinkhouse—a confusing and dispiriting change, it seems to me, since the flower honors the memory of a lady who was a well-known daffodil breeder. To her and her husband, whose surname is pronounced like the name of the Greek god of wine, we owe many of the red-cupped daffodils from which, presumably, the pink varieties were developed.

Soon it will be time to turn one's thoughts to growing flowers indoors. Because of a news item I've just caught up with, the prospect is sobering. The *IFAP News*, journal of the International Federation of Agricultural Producers, reports in its issue of last December that for the past eight years an Indian botanist, Dr. T. C. N. Singh, has been experimenting with growing plants to music. We all know that music broadcast in barns can make cows give down, but now Dr. Singh thinks he has proved that music broadcast in fields can make plants grow tall. He says that tapioca and sweet potatoes growing to the sound of recorded music showed a gain in yield of forty per cent, that rice plants exposed to Indian classical music increased their growth by fifty per cent, and that tobacco plants produced fifty per cent more when violin music was broadcast over the fields. His theory is that the sound waves agitate the plant cells in such a way that growth is accelerated. Now, as it happens, the worst trouble I am having with house plants is that they grow too fast. Our house is low-studded, and I have been anxiously eying an angel-wing coccinea begonia that I raised from a slip given me by a friend; at the moment it is reaching hard for the ceiling. A particularly handsome rubber plant that has been with us for two years has even fewer feet to go. Naturally, I can't help feeling that the sound waves in this house, which are considerable—the continuous chatter of voices, the scratchy strains of music all day and all night from several radios and a television set, and, less frequently, the pleasanter emanations from the piano and the record-player—are agitating my plants and turning them into giants. Dr. Singh appears to have uncovered a wholly new house-plant trouble. Nobody in my house wants to give up music. One answer to the begonia, of course, would be to eat it. The citizens of Paris ate their begonias as a substitute for spinach during the siege of 1871. But I doubt if I can face eating a rubber plant.

6

Green Thoughts in a Green Shade

Water lilies. From Breck's, 1908

March 11, 1961

I have read somewhere that no Japanese child will instinctively pick a flower, not even a very young child attracted by its bright color, because the sacredness of flowers is so deeply imbued in the culture of Japan that its children understand the blossoms are there to look at, not to pluck. Be that as it may, my observation is that Occidental children do have this instinctive desire, and I feel certain that almost every American must have a favorite childhood memory of picking flowers—dandelions on a lawn, perhaps, or daisies and buttercups in a meadow, trailing arbutus on a cold New England hillside in spring, a bunch of sweet peas in a hot July garden after admonishments from an adult to cut the stems *long*, or, when one had reached the age of discretion and could be trusted to choose the right rose and cut its stem correctly, a rosebud for the breakfast table. All these examples come from my own recollections of the simple pleasure of gathering flowers, but none of them quite equals my memories of the pure happiness of picking water lilies on a New Hampshire lake. The lake was Chocorua, and picking water lilies was not an unusual event for my next-older sister and me. We spent the best summers of our girlhood on, or in, this lake, and we picked the lilies in the early morning, paddling to the head of the lake, where the water was calm at the foot of the mountain and the sun had just begun to open the white stars of the lilies. The stern paddle had to know precisely how to approach a lily, stem first, getting near enough so the girl in the bow could plunge her arm straight down into the cool water and break off the rubbery stem, at least a foot under the surface, without leaning too far overboard. It took judgment to select the three or four freshest flowers and the shapeliest lily pad to go with them, and it took skill not to upset the canoe. Once the dripping blossoms were gathered and placed in the shade of the bow seat, we paddled home while their heavenly fragrance mounted all around us. I know now that their lovely Latin name was *Nymphaea odorata*, but at the time I knew only that they were the common pond lily of northeast America.

It is no wonder, then, that for years I have had designs on the lilyless pond that lies in the center of our pasture, here in Maine, where I live and from where I write. It is a small, heart-shaped pond, deep enough, I calculate, to grow hardy water lilies that will winter under the ice without freezing their roots. The only problem is the cows that drink at the pond and trample its grassy borders, but I doubt that they would wade in far enough to eat a small clump of lilies planted in the center of the pond and contained in a sunken box. The experiment would be worth trying. I could never attempt wild flag or the other shallow-water plants on the margins of the pond, for the cattle would devour them or tread them down, but a clump or two of lilies in the very center of our one small body of fresh water would satisfy me.

This is why I have been studying the catalogues of the water-lily growers. They make fascinating reading, since a water-garden specialist has to have more tricks to his trade than most nurserymen. Water lilies and lotuses are apt to be his primary horticultural crops, but with them he usually offers a list of floating plants, and shallow-water or bog plants to ornament the margins of the pool, and oxygenating plants to aerate it, and goldfish and scavenger snails to keep it clean, and frogs to eat the mosquitoes it will bring. Most growers also provide the pools, made of wood or fibreglass or steel, in which to plant the lilies, and water pumps, and underwater lights to illuminate a pool at night, and food for the goldfish, and remedies for fungus, algae, and goodness knows what else. This all sounds formidable, but I suspect that growing water lilies need not be complex if one's aims are modest. If one has a natural pond or a brook with a pool, the frills are unnecessary. Our own pond is plentifully supplied with bullfrogs, so I don't have to buy them at two to three dollars a pair, or tadpoles at a dollar a dozen. And the instructions for installing small artificial pools are so clear and simple in all the catalogues that even this task does not sound too difficult.

The five catalogues I have before me have taught me that there are day-blooming hardy water lilies—white, yellow, red, pink, blue, purple, and a whole range of colors in between—and tropical lilies, which are both day- and night-blooming, in just as wide a spectrum of color. These exotics are more floriferous and showy but on the whole less fragrant than the hardy varieties, and most of them will not winter over in a cool climate. In the North they had best be regarded as annuals. If I lived in the South or in California I would certainly want to grow a few, though, and even in the North two or three tropicals could be set out each spring for a moderate cost. One need only read the handsome catalogues of the Three Springs Fisheries (Lilypons, Maryland) and of the Van Ness Water Gardens (Upland, California) to be convinced. Three Springs is a father-and-son enterprise that has been in business since 1917. It started out as a farm and fishery, but there are now twenty-five acres of ponds for aquatic plants, set in the midst of more than three hundred rolling acres ten miles south of Frederick. We are told that Miss Lily Pons has adopted the post office as the place from which to mail her Christmas cards. Despite the punning name, which I can't admire, the beautiful black-and-white and color photographs of the catalogue make me long to visit Three Springs Fisheries and see its seventy-five varieties of hardy lilies, its twenty-five varieties of tropicals, and its lotuses and other water plants in bloom. To me the loveliest of the color portraits are of Gladstone, a hardy white lily; of Blue Beauty, a day-blooming tropical; and of the lotus Three Springs calls, not too correctly, *Nelumbo album striatum.* This sort of Nelumbo, the so-called Egyptian lotus, is one of our links with prehistory. A fossil lotus seventy million years old was recently found in eastern Asia, and in Washington, D.C., at the Kenilworth Aquatic Gardens, pink Manchurian lotuses are growing which were raised from seed that was a thousand years old when it was planted. One needs plenty of space for lotuses, which have leaves from two to three feet in diameter and flower stalks that grow to a height of from two to nine feet. (Unlike water lilies, lotuses bloom above the water.) Let go its own way, Nelumbo will create a small jungle, if that happens to be what you want. We have a native American lotus, of course, with yellow blossoms—the water chinquapin—and Three Springs grows and sells it, as well as ten lotuses from China, Japan, and other lands. The Van Ness catalogue offers far fewer varieties of both lilies and lotuses, but with its emphasis on warm-climate plants—small ones like water poppy and water hawthorn (which it spells "hawthorne"), as well as tropical lilies—it will be of special interest to those living in the Deep South or on the West Coast. Its cultural advice and planting diagrams are particularly helpful. Nearer New York City are three other firms: Slocum Water Gardens, of Binghamton, New York; William Tricker, Inc., of Saddle River, New Jersey (and also of Independence, Ohio, near Cleveland); and S. Scherer & Sons, of Northport, Long Island. Slocum has nine acres of display pools in Binghamton as well as ten ponds at Marathon, New York. Perry Slocum, the proprietor, bred the first patented hardy water lily, Pearl of the Pool, which in its picture is a very pretty pale pink. The Slocum catalogue lists four lotuses, which it claims are hardy and easy to grow; it is the only catalogue to lay much emphasis on the lovely scent of the lotus, which it says is the most fragrant of the water lilies. Tricker is one of the biggest growers of aquatics, but it lists fewer varieties of lilies and lotuses than Three Springs, and the catalogue is far from handsome, what with poor photographs, crude color, and unalluring typography. It is, however, comprehensive, and proffers all the aids and gadgets and fish and pools a water gardener could possibly need. It also has a long and interesting list of shallow-water, bog, and rock plants, both hardy and tropical. If one is after something truly gigantic, Tricker has it in an aquatic plant called a victoria (*Victoria trickeri*, to be exact, said to be hardy as far north as Cleveland), which has flowers eighteen inches in diameter and leaf pads six feet broad, strong enough to support a girl —a pretty one, I trust. (I'm surprised that the fashion magazines in their search for exotic backgrounds have yet to photograph a model wearing a high-style sports outfit while standing on the six-foot pad of a victoria in the center of a pond.)

The Scherer catalogue is a more modest affair than the others, but I like it. This Long Island nursery has been in business for fifty years and is the largest grower of aquatic plants on the Island; it is open to visitors weekdays, and is also open Sundays, holidays, and evenings in April, May, and June. This year Scherer has added to its tropical-lily list, it has fibre-glass pools in seven shapes, and with the help of its booklet one could install a pool and a charming water garden.

If a pool and water plants are too much effort, any one of these five catalogues is useful in another way. In each of their listings are all sorts of plants good for naturalizing and revivifying a dismal swamp or bog—wildings like cattails, marsh marigolds, the wild irises (yellow and blue), and, for running brooks, watercress.

From December to March, there are for many of us three gardens—the garden outdoors, the garden of pots and bowls in the house, and the garden of the mind's eye, whose midwinter vision, whether watery (like mine of the lily pond) or bedded in firm soil, is so brilliantly colored by the arrival soon after Christmas of the spring garden catalogues. This year my outdoor garden went under its blanket of snow the first week in December, and I have often been housebound. Therefore, I have had to enjoy the winter garden vicariously, with the help of books. The best for this purpose I've found is Elizabeth Lawrence's new one, *Gardens in Winter* (Harper), which has allowed me to share the delights of the author's garden in Charlotte, North Carolina, as well as the gardens and woods she knows from her wide reading. Miss Lawrence is a classicist, and can cite Virgil and the English poets as freely as she does Gertrude Jekyll and Jane Loudon. In this volume, she leads us most often to the English garden of Canon Henry Ellacombe, from whose two books *In a Gloucestershire Garden* (1896) and *In My Vicarage Garden* (1902) she often quotes. She also takes us to Walden Pond and Concord, in the winter sections of Thoreau's Journal. In North Carolina, Miss Lawrence says, there are two springs—the first in autumn, just after the killing frosts, and then the true spring, which starts on St. Valentine's Day—but she reminds us that it was Thoreau who wrote, "All the year is a spring," and her book seems to prove this to be true, even in the North. She has collected winter-garden notes and flowering dates from her network of correspondents all over the United States, and shares these with her readers. Though we Northerners will envy her her iris and camellias in November, her roses and hardy cyclamens in December, and her violets and hoopskirt daffodils in January, she shows us that all winter, even in the most frigid and unlikely spots, there are flowers or shrubs in bloom or, at the very least, in fruit, if we look for them carefully. (It is possible that on the sub-zero day in February on which I happen to be writing this paragraph, my Christmas rose, *Helleborus niger*, may be putting out its blossoms under the snowdrift that buries it; I poked and kicked at the drift a while ago, but it was too hard-packed and icy for me to uncover the flowers without damage to the plant.) It would be a mistake to give the impression that *Gardens in Winter* is merely a gardener's appreciation of the winter world outdoors. Miss Lawrence is a landscape architect and a skilled and discriminating horticulturist, and her book is also a practical guide to suitable and interesting trees, shrubs, plants, and bulbs for a winter garden. Sensibly, she includes in an appendix a list of plant sources and a list of winter reading, both technical and literary, for a gardener. I like the book most, though, because it quickens a country dweller's observation and stretches a gardener's imagination. The illustrations, pen-and-ink line drawings, are by Caroline Dormon, and they are as delicate and precise as the author's prose. Even the red dust jacket, with its lovely photograph of a clump of snowdrops flowering in snow, is an aesthetic pleasure.

Like Elizabeth Lawrence, I have in the past year been rereading the Countess von Arnim, the anonymous author of *Elizabeth and Her German Garden*, a book that was wildly popular with gardening ladies in the late nineties and early nineteen-hundreds. I tried it when I was a child, just to see why my aunts talked about it so much, but soon gave up. I disliked the Countess then, and I fear that I still do, in some of her moods. Her style is too ecstatic to suit this age, and she is alternately too coy and too heavily ironic to suit me, especially in her references to the Count, whom she invariably refers to as the Man of Wrath; her disagreeable comments on him and on her guests, whom she nearly always detested, make me uncomfortable. In fact, I now believe that they eventually made Elizabeth herself uncomfortable, for in her second book, *The Solitary Summer*, which I finally read this year, she seems to be trying to make amends for the misfiring of her irony and to show that hers was a happy marriage after all. The two books have a contagious enthusiasm for northland gardens and woods in both winter and summer, and their period pictures of life on a forbidding North German estate, worked by small armies of Russian serfs, and of the strict rules that governed the deportment of the mistress of the big house, are interesting. In her rebellion against this life, Elizabeth becomes a sympathetic character. Riding with her husband one day, she stopped to watch a group of women workers in the fields. One of them, arriving late, took up a spade and began to dig, and the overseer remarked that she had just been back to her house to have her baby. "Poor, *poor* woman!" exclaimed Elizabeth, and got into a wrangle with her husband about the sad lot of the women and whether it was right for the peasants to beat their wives. (He thought it entirely proper.) Poor, poor Elizabeth! She was not even permitted to plant her own flowers—unsuitable work for a countess—but, instead, fumed over a succession of inept gardeners:

I wish with all my heart that I were a man, for of course the first thing I would do would be to buy a spade and go and garden, and then I should have the delight of doing everything for my flowers with my own hands and need not waste time explaining what I want done to someone else. It is dull work giving orders and trying to describe the bright visions of my brain to a person who has no visions and no brains, and who thinks a yellow bed should be calceolarias edged with blue.

I see now why my New England aunts, who also disapproved of calceolarias, who cultivated their own gardens, and who had strong opinions on Rights for Women, were titillated by this book.

The meticulous art of flower painting had its golden age between 1760 and 1860, and its history parallels the history of horticulture in Europe and England, for as botanical explorers brought new species from the far corners of the world, the artists were apt to record them. Flowers have been painted from the beginning of time, and some of the earliest great botanical painters were the medieval monks who illuminated the borders of their missals with those exact and delicate portraits of flowers that Sacheverell Sitwell has called "miracles of human taste and skill." Mr. Sitwell and Wilfred Blunt are the editors of *Great Flower Books, 1700–1900*, a spectacular work that has also been part of my winter reading this year—that is, when I had the stamina to cope with this tremendous folio volume, nineteen and a half by thirteen and a quarter inches, and weighing eight pounds. Its handsome type is in general so large, and occasionally, in the legends on the thirty-six flower plates, so small, that this farsighted, astigmatic reader had to go at the book armed with reading glasses and a magnifying glass, but often found it better to wear no glasses at all. There is a reason, I suppose, for this outsize format; the venerable engravings from which these illustrations were taken were often very big, and much all-important detail would have been lost in reduction. The three dozen plates were printed for William Collins, the London publisher, in Amsterdam, the majority in eight-color photolitho-offset, and they are true rarities, not only because of the perfection of their reproduction but because many of the original flower engravings are hidden away in private collections. Mr. Sitwell contributes an enthusiastic if ornate essay on "The Romance of the Flower Book," and Mr. Blunt a more helpful one, on the history of flower painting. The botanical notes on the flowers depicted, and the introduction to the bibliography, which lists the major examples of this specialized art, are by Patrick M. Synge.

This noble book is naturally one to look at more than to read. The charm of the pictures, impossible to convey in words, is of course in part the charm of the flowers themselves, enhanced by the individual styles of highly skilled artists. When my Christmas rose was lost in a snowbank, I found it delightful to study the portrait of the Christmas rose engraved on copper by Nicholas Robert for Louis XIV. When my bird-of-paradise plant brought out its astonishing blossoms last October, I liked being able to study its likeness done by a great German draftsman, Franz Bauer, in 1818. And when I read in a catalogue of William Tricker's huge hardy victoria, *Victoria trickeri*, I could see its tender ancestor *Victoria regia*, as drawn in England by W. Fitch, in 1851. Mr. Synge's notes on this engraving say that *Victoria regia* is probably the most spectacular aquatic plant in the world, and he adds that when it first flowered, on November 2, 1849, in the Duke of Devonshire's collection at Chatsworth, it created a sensation. Perhaps the most enchanting pictures in the book come from France—Prévost and Redouté—but my favorite of all is by an Englishman, T. Baxter, who in 1806 painted a bold portrait of Bijbloemen and Bizarre "broken" tulips, in all their highly colored striations, against blue sky and white clouds. *Great Flower Books* has not yet been published in this country; and perhaps it never will be, because of the cost of producing a volume like this one; "great" flower books have usually come about at the behest of a rich monarch or a wealthy patron of the arts. The publication dates on the title pages of the old books were often years earlier than the actual publication dates, thanks both to financial exigencies and to the great care needed in the process of production. The English edition of *Great Flower Books* was small, but copies can presumably still be obtained from Collins direct, or from an American garden-book specialist.

I have further reasons for my interest in the Baxter tulip plate. Patrick Synge's comment on the picture includes these words:

> These are forms of tulips called Bijbloemens or Bizarres. The streaking is now known to be caused by a virus but at the period of the plate [1806] such tulips were far more popular than self-coloured forms and commanded very considerable prices. [In another place he writes] . . . the seventeenth- and eighteenth-century tulips seemed to be able to live with the virus and the plates are evidence of their vigor.

I quote these sentences in amplification because of my praise of "broken" tulips in these pages last September, when I failed to state that their lovely colorings are caused by a virus. I believe I knew this vaguely, but I thought little of it, because I did not understand that the bulbs are still infected. Now that I know better, I still think little of it, but it is only fair to say here that there is a Department of Agriculture bulletin on spring bulbs (Leaflet No. 439), which issues this warning:

> Many dealers sell Rembrandt, or "broken" tulips. These bulbs are infected with a virus disease that gives flowers a "broken" (striped, blotched, or mottled) appearance. Virus from these diseased bulbs will infect healthy tulips and lilies that are planted close to them. Diseased plants get smaller each year, die in 3 to 5 years. If you want to grow healthy tulips, you must keep "broken" tulips away from the healthy ones.

This indictment of the rare tulips that in Holland caused the great tulip mania has led me to a lot of investigation and correspondence. How, I wondered, could the Dutch and English growers have bred and sold broken tulips (I am tired of those quotation marks around "broken") and sold them by *named* varieties for centuries if the bulbs were still diseased? How could the characteristic markings that gave them their names have persisted so long? After correspondence with the government authorities (in particular, Mr. S. L. Emsweller, the head of the Department of Agriculture's Ornamentals Division), with horticultural authorities such as Dr. John C. Wister, of Swarthmore, and with two of the Dutch growers themselves, I found that all roads led to Dr. E. van Slogteren, of the Flower Bulb Research Laboratory, in Lisse, Holland, who everyone seems to agree is the greatest living authority on tulips. I was told that what he said about virus in tulips would be the facts. Luckily for me, Dr. van Slogteren is also an enthusiast of broken tulips, and published an article on them in the 1960 *Daffodil and Tulip Year Book* of the Royal Horticultural Society. He begins:

> In the opinion of the writer the "broken tulips" are the most beautiful of all tulips. . . . It will astonish outsiders to learn that the beauty of the flowers is due to a virus disease, causing a breaking of the colours. Yet this virus disease is after all the oldest known virus disease in plants and to my knowledge the only plant disease apt to effect an enhancement of the value of the infected plant.

He goes on to say that the virus is carried chiefly by aphids—particularly the peach aphid, *Myzus persicae Salz.*, which attacks trees of the Prunus family—though he has found it can also be transmitted by using the scissors that have cut broken tulips to cut tulips that are free of the disease. For more than a year now, he tells us, the Dutch bulb growers have been required by law to plant their broken tulips a hundred feet away from the healthy ones. But this matter of virus, which I have had to oversimplify, turns out to be only part of the story. It seems, according to Dr. van Slogteren, that there are a number of Dutch tulips that *resemble* broken tulips in their parti-colored markings but that are entirely free of virus, and sometimes, because of the similarity, they are listed by the growers under the heading of "Broken Tulips." This makes it confusing to the amateur gardener. In this category of cultivars that are virus-free, Dr. van Slogteren lists Prince Carnaval, General de Wet, Cordell Hull, American Flag, and (probably) Holmes King. There are perhaps others. Mr. Degenaar de Jager, of Heiloo, Holland, and South Hamilton, Massachusetts, the president of the bulb-growing and importing firm of P. de Jager & Sons, adds several others to the list, including Zomerschoon, a cottage tulip, which has been under cultivation since 1620. But according to Dr. van Slogteren, Zomerschoon is a broken tulip. If it is diseased, it is certainly stalwart to have survived all these centuries, and this makes me take heed of Mr. de Jager's letter to me: "The danger is within limited scope. If it was not, the whole of our stocks would have been infected and subsequently lost, since we have been growing a collection [of broken tulips] for over forty years among our other tulips, but have never found a spreading of the virus to the other varieties." Perhaps Heiloo is fairly free of orchards and their aphid pests, for Dr. van Slogteren, who is pleading for the continued cultivation of these historic and beautiful flowers, says unequivocally that they may become a dangerous source of infection to their neighbors. He writes, "This is certainly one of the reasons why the majority of growers today have stopped growing broken tulips. *But did they not throw out the baby with the bathwater in doing so?*" I have added the italics, because I think that Dr. van Slogteren is concerned, as I am, lest growers like P. de Jager & Sons and the others who have had the courage not to throw out the baby should be intimidated into doing so eventually, with the result that these beautiful strains would be lost.

As I said, I wrote in many directions in my distress, for fear that my carelessness and my enthusiasm for the Bijbloemens, Bizarres, and Rembrandts may have led readers into spreading virus in their gardens. My most encouraging reply came from the redoubtable Elizabeth Lawrence: "As to broken tulips, I had read that they have a virus but it never occurred to me that that was a reason not to plant them. . . . I hate the idea of making gardening a science, and I have nothing to do with it. My garden is for pleasure, I don't spray, or do anything you're supposed to do."

Well, neither do I, and this year, for the first time, my tulip bed is planted with a mixture of healthy *and* broken tulips. I hold over tulip bulbs from year to year, which Miss Lawrence, in the South, does not, so I may be in for trouble, but I am not alarmed, for the reason that I have had virus in tulips and lilies before now, from bulbs supplied by dealers who do not raise broken tulips at all. It didn't cause any real trouble, because I pulled up and burned any sick-looking plants. (Dr. van Slogteren is careful to say that the clearly patterned broken tulips are safer than tulips that have less obvious symptoms yet nevertheless are infected.) And if broken tulips have been grown—and painted by artists—for centuries, why should I not grow them, too? However, though I don't pretend to offer advice on gardening, I felt it only fair to give the warning that if you are one of those who want to be absolutely safe, you should either buy the named varieties of parti-colored tulips whose color patterns have not been caused by a virus or else plant your broken tulips at some distance from your clear-colored healthy varieties and your lilies.

To this lengthy Department of Amplification I shall add one last note, relative to the villains in the piece. It came in a letter, just after last fall's election, from a gardening friend in Washington who is a Democrat. The recount was in process and the outcome in doubt. "I do not wish to spoil your day," he wrote, "but I have just learned that all aphids are female. The Kennedy thing is all right. . . ."

I like, when I can, to compare the bright, shiny mail-order catalogues of today and their realistic color photographs of "bigger and better" flowers and vegetables with the seed annuals put out by the same companies when they were young. The comparison throws an odd sidelight on American progress—if it is progress. To be sure, many of the old favorites have disappeared or altered their methods of business. Peter Henderson's catalogue and its familiar cover picture of an old farmer whose benign, smiling face and laden barrow of vegetables used to conjure up such reassuring promises of plenty when the seed book arrived each spring are now only a happy memory, as is, for New Yorkers, the Peter Henderson store on Cortlandt Street, full of good, pungent smells and walled with neatly labelled drawers of seed from which dignified, aproned clerks filled your order. (In 1935, *Fortune*, to the dismay of the sentimentalists, reported that the body of one of these clerks had served as the model for the figure of the old man on the cover, but that for the face the artist inappropriately chose the face of an actor.)

If Henderson is gone, the huge Ferry-Morse Seed Company, of Mountain View, California, which celebrated its hundredth anniversary in 1956, is still going strong, although greatly changed in its methods since the days when it was D. M. Ferry & Co., of Detroit, and sold its seed by mail-order catalogue. Ferry-Morse issued its last catalogue in

1957, and almost from the start this company has distributed its seed at least in part on a commission basis, displaying the packets in racks placed in markets, hardware stores, and other retail outlets. Indeed, the founder of the company, Dexter Mason Ferry, may be said to have introduced the idea of self-service in stores (an idea I wish had never been thought up), for it was he who invented the so-called "commission box," or seed rack. For generations, though, the company was in close touch with its customers by mail. One of my precious possessions is the Ferry "Seed Annual" for 1895. Its cover, in soft pastel tones, is a reproduction of a charming water color of three Gibson girls in a grassy field of wild flowers. One girl, wearing a white dress and a stiff sailor hat, stands knee-deep in the grass, gazing into the distance; one, in beige, has sat down to rest on a bank of flowers while she shields her face from the disfiguring sun with a white ruffled parasol; the third, in pink, is approaching her friends, holding up in the folds of her long skirt the bouquet of field flowers she has gathered. *Not a single flower in the entire picture is recognizable*, a fact I find engagingly non-commercial. It was enough for the Mr. Ferry of 1895 to suggest that pretty girls liked pretty flowers, whether or not the flowers were raised from Ferry seeds. The only publications Ferry-Morse puts out today are a planting chart, and three small leaflets that give, respectively, elementary advice on garden culture, on how to make your meals tastier with herbs, and on "Fixing Flowers to Please Your Family." The cover of this last is a color photograph of a gray-haired, well-permanented matron arranging flowers in her modern kitchen, following, I've no doubt, the Ferry-Morse rules for taste, or perhaps experimenting with the leaflet's suggestions on how to have "Fun with Forms." Even flower arrangement, it seems, is now a mass commercial movement. The whole sad business is a far cry from the romantic cover and the interesting catalogue of 1895, which offered eighty-eight pages of vegetables and flowers. The flower list described thirty-seven *named* varieties of sweet peas and the seeds of many flowers now almost unknown to commerce, and of course most of the flowers we grow today. But if there is small call for the Chilean gloryflower (*Calampelis scabra*) or *Cedronella cana*, a balm from the Canary Islands—to name only a couple of varieties sold in 1895—Ferry-Morse has its substitutes, for the firm was also one of the pioneers of scientific seed breeding. For example, its sweet pea Blanche Ferry, pictured in color in the 1895 catalogue, is the forebear of all our early-flowering sweet peas. The Morse half of the firm was probably responsible for the sweet pea, since long before the merger of the firms, Morses were breeding flowers for Ferrys. It happened this way: Charles Copeland Morse, a lad from Thomaston, Maine, arrived in California in 1862, sailing before the mast, and decided to stay there and raise seeds. By 1896, his company was the chief producer of seeds on the Pacific Coast, and his son became the West Coast's leader in improving old strains of seeds and developing new ones. The merger of the two companies was made in 1930, and the firm is still hard at this business of plant breeding. It is, for instance, responsible for our newest sweet alyssum, Rosie O'Day, which this year won an All-America Award and which is touted in almost all the catalogues of the seed houses that still bother to publish catalogues. Rosie O'Day is a deep, unfading rose-pink, with a flat habit of growth, and in her color pictures she is very pretty.

There is at least one seed company that was *founded* on the principle of the commission box—the Mandeville & King Co., of Rochester, New York, which for the past eighty-five years has specialized in flower seeds and sold them in retail stores. It, too, issues no catalogues, and I find it a pity for those of us who do not happen to live in the region where

Mandeville seeds are distributed. I have never been in a position to buy one of its packets off a seed rack, but an unknown friend has sent me one of them—a packet of sweet-lavender seed. The legend on the little envelope gives full cultural instructions, preceded by the following sentences:

Lavender is grown for its fragrant flowers. They are used for making sachets and perfumes. The flowers are dried and laid with the linen to make it fragrant. The Kabyle women of North Africa think it protects them from maltreatment by their husbands. Most American women know better.

If all Mandeville's packets read as well as this one, how pleasant it would be if the firm would issue a catalogue full of just such useful and provocative information and would sell its wares by mail all over the country.

Happily, most of the old mail-order seed houses are going as strong as ever today. Before me as I write are Burpee's, Harris's, Breck's, and Park's catalogues, each of them as full of life and promise and happy surprise as they were for our grandparents. I have made a discovery about the W. Atlee Burpee Company. The firm issues *two* retail catalogues; one is slightly smaller and much less comely than the other, but it has the same seed lists and black-and-white photographs, and the same front and back covers in color. It lacks, though, the twenty-four good color pages that the second, far better-looking catalogue has, and is printed on poorer paper, which makes the photographs less sharp. The explanation is simple. The Grade A catalogue goes to the firm's steady customers, the smaller one to those who have not put in an order for some time and are, in Burpee's eyes, forlorn hopes, worthy only of an economical, faint-chance seed book. The plan makes sense, and Burpee, by following it, is far more generous than many firms, which drop you from their lists whenever you fail to order. Burpee presents some new marigolds this year, two of them named Alaska and Hawaii. (What would spring be without a new Burpee marigold?) Hawaii is bright orange; Alaska is a pale primrose, yet it is the nearest to white of all the *big* marigolds. There is, though, a new and smaller marigold that is almost pure white; its name is Whitey.

The George W. Park Seed Company is one of the last of the seed companies to sell sweet peas by name. It still offers a few name varieties. For a really big assortment of named sweet peas, though, you will have to turn to the exhaustive catalogue of Harry E. Saier, who lists nearly fifty varieties by name. The fat booklet is well worth sending for, if you want unusual flowers, because the energetic Mr. Saier, who is seventy-three years old, collects his seed from all over the world; his listings, astonishingly, include approximately eighteen thousand kinds of seed. Here you can even find Ferry's 1895 glory-flower and Canary balm, should you want to grow them, but you will have to look up glory-flower not under its common name or under *Calampelis scabra* but under its corrected Latin name, *Eccremocarpus*. Mr. Saier has little patience with familiar names. The cover of his catalogue this year carries the words "Fifty Years," and it is Mr. Saier himself who has built up his great collection over half a century.

The Joseph Harris Company this year adds to its last year's prize-winning Rocket snapdragons a new strain of shorter ones, called Frontier snapdragons, which are also F_1 hybrids. They grow to only two feet and produce as many as fifteen spikes to a plant. I like snapdragons in part for their stately height, but these short ones, planted in front of the tall ones, should make a pretty flower bed. Harris has this year a new dwarf salvia called Red Pillar, which in its pictures is, I must admit, the best-looking red salvia I've seen, although why anyone ever wants to grow salvia unless it is the blue perennial variety I don't know. The plant breeders this year seem to have turned their attention to improving flowers I have never been able to admire. I had assumed that red salvia, cannas, and coleus were Victorian plants we'd never have to bother about again, and believed they would soon disappear from our garden landscape, together with the last of the iron deer. Yet here is Harris with its salvia, and the George W. Park Seed Company with its brand-new coleus, named Coleus chartreuse, and Breck's is featuring still another salvia, the dwarf Scarlet Pygmy. Even the handsome new book of Wayside Gardens, which is bigger and more fascinating to study than ever, to my amazement devotes three pages to its new pastel cannas and its hybrid dwarf cannas from Germany. "Gone," writes the cataloguer, "are the hard, offensive colors of the old Canna. Gone also are the small, spindly blooms." Quite true, but what I did not like in the first place about cannas, coleus, and red salvia was not so much their various colors as the texture and design of their flowers and leaves. The shape and substance of a flower seem to me to determine its personality quite as much as its coloring or its fragrance. The flower heads of the improved cannas, for instance, are still just oval blobs of color. However, even I can visualize a spot where the new chartreuse coleus might make a lovely patch of fresh yellow-green, where the soft tones of the pastel cannas might be effective, and where the red fire of the salvias might be cheerful.

Last fall I gave away all my gladiolus bulbs, deciding that since I cared so little for them I could not endure one more winter of drying them out and hanging them in the cellar.

I bought the first ones years ago, chiefly because doing so gave me an excuse to pay a call on our First Selectman, who grew gladioli as a sideline. He is long since dead, yet for years, perhaps partly to remind me of my pleasant visits in his kitchen, I have grown a row of gladioli in the vegetable garden, not knowing where else to put these awkward flowers, which take up so much space and do not combine well with other flowers in a border or bed. Despite my neglect, they performed faithfully each summer, giving me enormous stalks of white, cream, lavender, yellow, salmon, and pink flowers, which I dutifully cut and brought into the house when I had nothing better to pick. Growing, gladioli seem to me ugly, poking their great spikes out of the ground at queer angles, and knocked down by every heavy rain, and I find them ungainly to arrange as cut flowers unless I have a huge vase and a big, stately room, which I do not have in this old farmhouse. When they are cut short and arranged in a shallow container in the modern pseudo-Japanese style, I do not enjoy them, either. The sad truth is that I have never been able really to like these showy, impersonal, unfragrant blooms, which remind me of my few hot-weather sessions in a hospital, since in midsummer florists seem to make them the mainstay of their hospital deliveries. All this is doubtless my fault; several of my friends are able to arrange them in lovely bouquets. I have to admit that some of their colorings are lovely, and today, after studying the handsome color and black-and-white photographs in the new catalogue of Champlain View Gardens (Burlington, Vermont), I almost regret my stern decision never to grow them again. The catalogue, I am sorry to say, is titled "Gove's Glad Book," and the foreword is signed "Elmer Gove, Apostle of Beauty." (Half my prejudice may lie in this flower's uneuphonious name, with its awkward plural, and especially in its horrid nickname. I can *not* be glad about glads.) Nevertheless, to the countless numbers of less prejudiced gardeners I commend Mr. Gove's good-looking and complete catalogue, and I call their attention to two lovely new blue gladioli named China Blue and Blue Mist and two luscious lavenders named Blue Lilac and Orchid Charm. Mr. Gove has everything a fancier could want—thirty color classifications, exotic hybrids, miniatures, even a half-dozen fragrant varieties. He appears to be a trifle doubtful about these last, though, for he writes that some people can't seem to get their fragrance, and that if they can't they are not to complain to him, for he does not list any variety as fragrant unless it really is. I can hardly imagine what a scented gladiolus would smell like.

"Every year, in November," writes Maeterlinck, "at the season that follows the hour of the dead, the crowning and majestic hour of autumn, reverently I go to visit the chrysanthemums. . . . They are, indeed, the most universal, the most diverse of flowers." Maeterlinck is, to me, next to intolerable when he writes on gardens—exalted, flowery, and overblown—but I do happen to agree with him on chrysanthemums, and daily in autumn, I, too, visit reverently each chrysanthemum plant that has survived the frosts of October's Allhallows Eve, "the hour of the dead." I love chrysanthemums for their resistance to frost, their diversity of form, their strong, pungent foliage, and their great range of cheerful yet subtle autumn colors. In Maine it is difficult to bring them through a cold winter, and we always lose a few, but I prefer to do that rather than go to the trouble of wintering them in cold frames, as many do in severe climates. Besides, a new chrysanthemum plant, set out in spring, will, if pinched back to cause branching, make a fine show by the following October.

The Bristol Nurseries (Bristol, Connecticut) specializes in chrysanthemums and has comprehensive lists of all the various types—the Decoratives, the Spoons, the Pompons, the Spiders or Rayonnantes, the low-growing Cushions, and the Football chrysanthemums of the florist shops. The catalogue is illustrated with bright color photographs, but because the 1961 edition is not yet out, I cannot describe it in detail. I advise you to send for it and order some plants, as I did a year ago. Mine were good specimens and grew well.

Lamb Nurseries, too, has an excellent list of American chrysanthemums, but it makes a specialty of the early-flowering English chrysanthemums, which are the largest of the outdoor varieties, with flowers five to six inches across and petals reflexed, in-curved, or shaggy. These English varieties, Lamb says, will bloom from July until the frosts come, but I gather that they are not as hardy or frost-resistant as the other types. They should be useful, though, where the cold comes extra early.

I myself prefer the smaller kinds, the Buttons and the Pompons, and the medium-large types with looser petals. The Putney Nursery, of Putney, Vermont, has a good small assortment, and I liked its Abundance, a bronze with reddish eyes; Football Bronze, a handsome in-curving type; Lee Powell, a China gold; and Lavender Lady, a silvery pink-lavender. Most of my chrysanthemums, though, came from Wayside Gardens. My Wayside favorites are the luminous Ashes of Roses, of which I have written before, and Bronze Cactus, which is a *pink* bronze. Put these two near Ancaster Peach, and the peach-pink Mrs. Pierre S. du Pont III and the others named above, and, unlikely as it sounds, they blend as well as a row of frost-touched maples. We have three maples between our house and the road, one of which always turns a pink red, another a pale yellow, and the third a chartreuse green, and one sees them against the scarlet of the swamp maples across the highway. They color just when our chrysanthemums are in flower, and for me both the flowers and the trees spell the best days of autumn.

This year I shall postpone until autumn my description of the new roses and new rose catalogues; a few growers do not issue their complete catalogues until fall, anyway, and all of them emphasize the virtues of fall planting. However, a hasty inspection of the rose books I have at hand makes me aware of one happy change, which I cannot wait to celebrate. This is that the growers of modern roses more and more are taking pains to tell us about the fragrance of their roses or, by omission of any mention of scent, to imply its absence. For consistency on this point, I give top prize to Peterson & Dering, of Scappoose, Oregon, which has the most orderly rose catalogue I've seen yet. The roses are grouped by colors, their American Rose Society ratings are given, and each scented rose is graded conscientiously as very fragrant, fragrant, or slightly fragrant. It is notable, in this age of the non-fragrant rose, how many of Peterson & Dering's *are* sweetly scented. Melvin E. Wyant, too, has apparently been sniffing every rose he grows, and he goes so far as to describe the individual scents as his nose dictates them to him: Sutter's Gold has a "rich tea fragrance," Grande Duchesse Charlotte has a "slight carnation fragrance," Katherine T. Marshall is "fruity" and Joanna Hill "musky," Rex Anderson smells faintly of lemon, and so on. "Star Roses," the catalogue of the Conard-Pyle Company, often makes the same sort of descriptive effort and in general is far more helpful on the matter of fragrance than it used to be. Will Tillotson's Roses and Joseph J. Kern, whose major interest is the sweet-smelling old-fashioned rose, have always noted fragrance. Jackson & Perkins, on the other hand, and perhaps still the majority of the rose growers, do not bother to mention scent except when it is present in the new roses they happen to be featuring. J. & P. customers are supposed to know without being told that the modern but not recent Chrysler Imperial and Sutter's Gold, which the company still sells, are notable for their perfume, but though this asset is doubtless taken for granted by the rose men themselves, many who are starting their first rose bed will not know it. Perhaps if the growers realized that roses described without mention of fragrance are usually assumed by amateurs to have none, they would take more care.

For a time, my belief that a rose without scent was only half a rose caused me to wonder whether I had an obsession about this matter, but recently I have come on a letter that makes me think that if I have, I am in good company. The letter was written in 1573 by the seventeenth Earl of Oxford to his friend Thomas Bedingfield:

> What doth avail the rose unless another took pleasure in the smell? Why should this tree be accounted better than that tree but for the goodness of his fruit? Why should this rose be better esteemed than that rose, unless in pleasantness of smell it far surpassed the other rose?

Why indeed? The seventeenth Earl of Oxford was a man of many gifts. He may or may not have written the plays of Shakespeare, but I do find the Earl a kindred spirit, and I trust that the rose men will pay heed to his words.

If it were practical, I would some year devote my entire spring report to the delights of the English garden catalogues, which send me, as a devotee of this form of reading matter, into an ever-green, ever-blooming, idealized garden that is compounded of many things both literary and actual: English gardens I have seen in Cornwall and the Cotswolds, in Buckinghamshire and Surrey; the Chelsea Flower Show, which I visited just before the Germans invaded the Lowlands in 1914; the soft, equable English climate, "annihilating all that's made to a green thought in a green shade;" the English garden books, which are usually so much better than ours; and, perhaps most of all, the landscapes and shrubberies and gardens and flowers of the English novelists and poets. Jane Austen, who lived in the age of the "improvers"—as landscape gardeners like Humphrey Repton and his colleagues were then known—is particularly good (despite Fanny Price and the headache she got picking roses for Aunt Norris) on great avenues of trees, formal prospects, ha-has, and shrubberies rather than on flowers, but it is to the poets, from Shakespeare to Wordsworth, that we owe our best pictures of English wild flowers and garden flowers. All this background of our English heritage gives a special aura to English garden catalogues, at least to American readers—an aura that probably does not exist for the British, to whom the English garden and the English landscape, whether literary or actual, are a matter of course.

There is, as an example, the forty-eight-page book of Allwood Brothers, Ltd., which is devoted entirely to carnations and pinks for the garden and greenhouse. I presume the British take it for granted, and we who do not may also have it by sending twenty-five cents to the nursery, at Hayward's Heath, Sussex. The Allwood garden pinks are obtainable in many American catalogues—Wayside's, for one—and whenever you see laced pinks or Allwoodii pinks or border pinks offered for sale, they probably came originally from Allwood seeds or Allwood plants. Yet the meagre offerings of the American catalogues can give you no idea of the variety Allwood has to offer in its own listings or of the charm of these aromatic old-fashioned flowers, which is so well shown by monochrome photographs in Allwood's unpretentious catalogue. We are not told by Wayside that the laced pinks have names like Charity, Faith, Felicity, Hope, and Verity, nor can we usually choose the colorings of our pinks in this country. The Allwoodii varieties are a hybrid between the old-fashioned garden pink and the perpetual flowering carnation. They bloom all summer, they are free-flowering, and every one of them is deliciously scented. *Allwoodii alpinus* is the race of dwarf, spicy, ever-flowering pinks for rock and wall gardens. Then there are the Border and Cottage and Picotee carnations, and the Village pinks, and the Bedding pinks, all for the garden—not to speak of the big carnations for cool greenhouses and, also for greenhouses, the perpetual-flowering Malmaisons, "remarkable for their outstanding perfume" and beloved of King Edward VII and Queen Alexandra. The wealth and variety positively make the head swirl. The wide range of colorings is equally surprising; until I read this catalogue I had no idea that common garden carnations may have apricot, crimson, helitotrope, scarlet, and yellow grounds as well as pink and white. Allwood's descriptions are uneffusive and to the point. Let us take that of the laced pink Hope, which won a Royal Horticultural Society award: "White ground, good crimson lacing, semi-double scented flowers. Free flowering." The named varieties are sold as plants, and unfortunately this year the plants may not be brought into the United States, even with an import license. The seeds need no license and always come in packets of mixed colors. It would be pleasant for Americans if Allwood would sell its seed by named varieties, at least until the embargo on plants is lifted.

From England, too, I have this week received a new recipe for the plant that is sick or dispirited. It comes from Mollie Panter-Downes. A friend of hers in Surrey was showing an ailing wisteria vine to a gardening acquaintance. "Oh," said the visitor, "all that wisteria wants is a nice rice pudding. They *love* them!" Accordingly, a rice pudding was cooked, well sugared, and laid round the feet of the vine, which promptly sat up, regained its tone, and is now full of health and pudding. There is probably a chemical explanation for this, but I would rather not know about it.

7
For the Recreation & Delight of the Inhabitants

Lawn mower. From Breck's, 1884

June 6, 1962

Home would not be home to me without a lawn, and if there are, as I've recently read, twenty-five million home lawns in the United States, at least fifty million other Americans must agree with me. Yet the more one considers the never-ending labor a lawn represents, the more astonishing it is that in every small city, suburb, town, and village, our patches (or wide expanses) of greensward are so lovingly tended year after year. The effort and expense of making a lawn—grading, topsoiling, rolling the ground, and sowing the seed—are minor compared to the week-by-week, year-by-year fatigues of keeping the grass fertilized, free of weeds, watered, and shorn. Why so many of us consider this green velvet ground cover essential to the beauty and tranquillity of our homes is a question to which I have never seen a satisfactory answer. The sociologists call the lawn a "status symbol," and it is, to a degree, since nearly everybody wants his bit of ground to look as neat and pretty as the next man's. Yet I think there is much more to it than that. We could have chosen, in the European or Oriental tradition, to pave or cobble our yards and garden areas, and modern landscape architects, even in humid England whose turf is our envy, are urging everyone to do just that. To me it would be a great loss.

Consider the many special delights a lawn affords: soft mattress for a creeping baby; worm hatchery for a robin; croquet or badminton court; baseball diamond; restful green perspectives leading the eye to a background of flower border, shrubs, or hedge; green shadows—"This lawn, a carpet all alive/With shadows flung from leaves"—as changing and as spellbinding as the waves of the sea, whether flecked with sunlight under trees of light foliage, like elm and locust, or deep, dark, solid shade, moving slowly as the tide, under maple and oak. This carpet! What pleasanter surface on which to walk, sit, lie, or even to read Tennyson?

> *Sweeter thy Voice, but every sound is sweet;*
> *Myriads of rivulets hurrying thro' the lawn,*
> *The moan of doves in immemorial elms,*
> *And murmuring of innumerable bees.*

The familiar lines are perhaps as apt for England today as when Tennyson wrote them, but our immemorial elms are fast dying, our bees, thanks to poison sprays, are now only too easily numbered, and for the same reason our doves may soon no longer moan or our robins hop about the lawn and cock their heads to listen for the good earthworms underneath the grass. It is a rare American lawn, too, through which rivulets hurry, and the weary American businessman (or, for that matter, the weary American plowman) must wend his way home after a hard day to turn on the hose and the sprinkler, and, more often than not, must also be prepared to give a part of his precious weekend leisure to cutting the grass.

Trying to find out what started it all, I went first to the big dictionaries. "Lawn," according to the Second Edition of Webster's Unabridged, is of Celtic origin and derives directly from the Middle English "*laund*" or Old French "*launde*," meaning "heath, moor." Webster's first definition, now archaic, is "an open space between woods; a glade." The Oxford English Dictionary gives as other early meanings of the word "a stretch of untilled land" and "an extent of grass-covered land," and cites the 1733 edition of Philip Miller's famous *Dictionary of Gardening* for the first use of the word in something approaching its modern sense: "a great Plain in a Park, or a spacious Plain adjoining to a noble Seat."

Those eighteenth-century spacious plains of England were kept cropped by sheep or cattle, as were the first great lawns of the United States. U. P. Hedrick, in his valuable and carefully documented *History of Horticulture in America to 1860*, not only speaks of the early manorial lawns, such as those at Mount Vernon and Monticello and the bluegrass lawns on the great estates in Kentucky and Tennessee, but tells us that even in the very early days of Colonial America there were grassy public commons, squares, and parks, and fine-turfed bowling greens. The game of bowls was brought to us by the Dutch settlers, and it remained a fashionable and popular out-of-door American pastime almost to the start of the Civil War, when croquet began to take its place. Every large estate had its bowling green—Washington placed his at Mount Vernon just north of his vegetable garden—and there were public greens on the village commons and in the yards of inns. Bowling Green Park, the first public park in New York City, was established in 1733, when the Common Council leased "a piece of Land lying at the lower end of Broadway fronting to the Fort," in order to enclose it and "to make a Bowling Green thereon, for the Beauty and Ornament of said Street as Well as for the Recreation & Delight of the Inhabitants of This City."

Recreation and Delight, but also gruelling hard labor, for the lawnmower was not invented until 1830. Where sheep and cattle proved too untidy—near a dwelling or in cities and towns—American and English lawns and bowling greens were sheared with scythes and sickles, and the grass cuttings were swept up with brooms. You may get an idea of what this sweeping alone entailed by reading John Claudius Loudon's precise directions for cleaning a lawn after a morning's mowing. They are quoted in full in Geoffrey Taylor's delightful little book *Some Nineteenth Century Gardeners*, published in England in 1951. Loudon, the one-armed genius of the early Victorian garden scene, and his wife, Jane, who was his amanuensis and his partner in gardening, are the subjects of Dr. Taylor's first two chapters. The husband was a naturalist, landscape painter, traveller, architect, writer, and editor. He founded the first garden magazine in the English language, and wrote many books, including his influential *Encyclopedia of Gardening* and his massive, eight-volume *Arboretum et Fruticetum Britannicum.* Primarily, though, he was a gifted gardener, and Dr. Taylor says that modern gardening may well be considered to begin with Loudon. He was, above all, a perfectionist. According to his plan for gathering grass cuttings, a man with a hay-rake is followed by a man with a nine-foot besom (a broom made of twigs), who is followed by a woman or boy with a four-foot besom, who is followed by a man with a grass cart and a boy with boards and another short besom for scooping and brushing up the cuttings that have been left in neat piles by the first three laborers. There is, as far as I know, no American edition of Dr. Taylor's book of portraits of the men who, like Loudon, most influenced gardening in the last century, but get hold of the little volume if you can, and its equally illuminating sequel, *The Victorian Flower Garden.* These books so charmed and informed me that I must have read them several times over in the past year. The facts are exact, since Dr. Taylor is the director of the Royal Botanic Gardens at Kew, and for good measure this distinguished horticulturist is blessed with a sense of humor and a graceful style. "Note on Nineteenth Century Lawns and Lawn-Mowing," the final chapter of the first of the books, tells us the story of the lawnmower—and a curious one it is.

An English engineer named Edwin Budding was employed by a textile factory where a spiral-blade machine was used for cutting the pile of certain fabrics. Seeing no reason that the principle could not be applied to a machine for cutting the pile of lawns, he worked in his spare hours to develop one. Thus his great idea, to which homeowners owe so much, was more an adaptation than an invention. When he completed his mower, he tried it out surreptitiously on his lawn, doubtless to the amazement of his neighbors, for he made his trial runs only after dark. The grass cutter produced a great clatter, but it worked so well that he patented it in 1830, and a year later "Mr. Budding's machine," as it was then generally called, was being manufactured by two companies. It was a cumbersome affair, but Dr. Taylor tells us that the head gardener at the London Zoo calculated that "with two men, one to draw and another to push, the new mower did as much work as six or eight men with scythes and brooms." By the eighteen-sixties, horse-drawn mowers were in use at Kew Gardens, but in the same decade the gardener of an English insane asylum, a resourceful man, found it easier not to bother with horses; he harnessed seven madmen to his machine. There is also a story that one Englishman hitched a camel to his Budding mower and then dragged both the animal and the machine around his lawn. In the early eighties, a Mr. R. Kirkham invented a mower propelled by a man on a tricycle, and a decade later a Mr. Sumner patented a one-and-a-half-ton steam mower. Neither of these contraptions caught on, so men went on pushing their lawnmowers until some years after the invention of the combustion engine. Appropriately, the first power mower was conceived in Detroit, when a Colonel Edwin George attached the gasoline-powered engine from his washing machine to his lawnmower. It did the trick, and in 1919 he established a company to make and sell his Moto-Mower. The rest is modern history, and our own happy memories of life on the lawn can supply the details.

There were no "noble Seats" or bowling greens in the suburb of Boston where I grew up—no cattle, no sheep, and, of course, no power mowers—but our lawn was kept impeccably shorn and green, at least to my young eyes. Indeed, so dear to my father's heart was "The Lawn," and so important a part did it play in my childhood, that I tend now to think of that modest, grassy slope on Hawthorn Road in the eighteenth-century capital letters of a Philip Miller or a New York Common Councilman. My father did not have much time to give to the grass himself; six days a week he left the house by eight, to go to his office in Boston, and he returned only a few minutes before seven. Nevertheless, six or so months of the year, he was up and out on the lawn before his early breakfast, to inspect the work of our handyman and to dig weeds out of the grass with his pearl-handled penknife. The sight of a dandelion in bloom in the grass so inflamed him that one bad year for dandelions he equipped his two younger daughters with jackknives and offered us ten cents a hundred dandelion roots, provided we cut them nice and deep. I can't remember earning a great deal of pocket money that way, and I think now that my father gave us the knives quite as much because he wanted companions for the game of mumblety-peg—one of the many games that can best be played on a lawn—as because he expected us to make any real headway on the weeds. He was an expert at this simple sport, and to this day I can still execute some of the throws he taught me—the knife placed, blade open, in various positions on the outstretched hand, the wrist kept supple to make a throw that will leave the knife standing erect, its blade imbedded firmly in the turf. (The most fun was the toss over the shoulder.)

It was William Hickey, our neighborhood handyman, who mowed our lawn each week with a hand mower, trimmed its borders with hand clippers, and set the whirligig sprinkler to work in times of drought. William's life, unlike that of his more celebrated namesake, the eighteenth-century rake and diarist, was no round of fun. I wonder now how he ever did so much for us and his other clients. For us, in summer, besides tending the lawn, he clipped the privet hedge, raked the driveway, and weeded the beds; in winter he stoked the cranky coal furnace, shovelled snow, and carried in wood for the open fireplaces; all year round he blacked our boots and, night and morning, lugged two hods of coal from cellar to kitchen, to fuel the big iron-and-brick stove. Every Monday he had to make another coal fire in the basement laundry stove, and every Saturday he washed our flea-ridden smooth-haired fox terrier with Knock-Em-Stiff, in one of the laundry tubs. (I presume he did just as much for our neighbors.) In early spring, William's lawn duties at our house were made a trifle lighter by the arrival of a pair of horses and a small army of men, whom my father hired to put the lawn in shape, dress the flower beds, and prune the shrubs. The horses were my special delight, because they wore short leather boots over their hoofs to prevent their iron shoes from cutting up the turf as they spread manure and dragged a heavy roller over the grass. One dreadful year, the whole lawn had to be remade by the same team of men and horses after an infestation of June bugs, which laid their eggs in the grass. We were away that summer, and the attack was not discovered until the eggs had hatched out into death-dealing white grubs.

That year was the only one we children did not as good as live on the lawn out of school hours, playing a great variety of games on the grass—croquet, baseball, mumblety-peg, ring toss, diabolo, battledore and shuttlecock (never by us called badminton), and, best of all, a game we called "statues." In our local version of this game, a soft carpet of grass was a true necessity, for the little girl who was "It" would seize each of the other girls in turn by the wrist and whirl her around and around until she was dizzy, and then, just as she flung her victim off, to stagger, and often to fall, on the grass, she would scream out an abstract noun—Grief, Anger, Grace,

Hatred, Love, or Defiance—to indicate the pose and expression the living statue must freeze into. The child who in the whirler's opinion took the most outlandish pose and assumed the most appropriate expression became the next "It." Statues was a very violent game, and it could never have been played in a stone- or brick-paved back yard. It occurs to me now that it may well have been thought up by a little Bostonian who, like me and my friends, had been compelled to paste up for her class in ancient history an album of Copley prints showing the classic Greek and Roman sculptures. Do Copley prints still figure in Boston education? I wonder. Do little girls still posture on the lawns of Hawthorn Road?

To come out of this green nostalgic haze, I should report that our lawn in Maine today is mowed once a week, except when there is a drought, by a noisy power mower, making the day hideous and causing me to remember wistfully the drowsy, soothing hum of William's hand-propelled machine. We have never used chemical weed killers and seldom have time to employ a jackknife, nor do we water the lawn. Steady mowing and fertilizing, plus resowing the poorer spots in spring, and again in late summer if we remember to, seem to keep the grass green and free enough of weeds. To be sure, a damp section under the shade of an oak is more mossy than grassy, and there are rapidly spreading patches of a leafy weed that I think is one of the hawkweeds, but moss stays green in the fogs of a Maine summer and the weedy patches are as green and almost as pleasant to walk on as the grass. We used to roll the turf each spring, but now we find it as good, if not better, without rolling. We catch the grass cuttings in a bag and add them to the compost heap, and at least some of them are returned to the lawn in the form of compost. (The preferred method, if you have the manpower, is to give the grass a light cut every four or five days and let the short clippings fall, to make a protective mulch.) We have no consistent lawn program, taking each year as it comes. White Flower Farm has one, though; it devotes a section of its 1962 "Plant Book" to lawn seed, and issues a sheet that it calls "A Lawn Transformation Program." I agree with most of the White Flower instructions, taking strong exception only to its advice on weed killers and to its firm recommendation of White Flower's own lawn seed, which is a mixture of eighty-five per cent fescue grass and fifteen per cent bent grass. We are not told what kind of fescue and what kind of bent grass make up the formula. Fescues are useful in dry soils and in shade, bent grass is the golf-course grass and needs much care, so where there is sun and an average soil I prefer a mixture that contains a preponderance of bluegrass. Every region and every lawn is different, of course, and lawn seed must be chosen to fit the conditions. *Hortus Second* contains an interesting entry on lawns and grasses, and the 1961 revised edition of *Taylor's Encyclopedia of Gardening* has an even longer essay, which includes more recent advice. The Taylor of the *Encyclopedia* is an American, Norman Taylor, who is not to be confused with Dr. Geoffrey Taylor, of Kew Gardens. Norman Taylor, formerly associated with the New York and Brooklyn botanical gardens, is known to gardeners chiefly for his many helpful garden books, of which this 1961 fourth edition of the *Encyclopedia* is his greatest achievement. It contains hundreds of articles by excellent authorities, including Mr. Taylor himself, on gardening techniques, landscape, design, and horticulture, and also describes over nine thousand plants, both wild and cultivated, now growing on this continent. *Hortus Second*, on the other hand, by the venerated Liberty Hyde Bailey, of Cornell, is a dictionary, "designed," as its preface says, "to account for all the species and botanical varieties of plants in cultivation in continental United States and Canada in the decade ending midyear 1940, together with brief directions on uses, propagation, and cultivation." I have no idea how many species and varieties of plants this would add up to, but many, many more than Mr. Taylor's nine thousand. The two books neatly supplement each other, and I use them constantly. Reference books of this sort are usually more reliable than anything put out by a nursery or seed company, and they will tell you what kind of seed and what kind of program are best for your region and your soil.

This is why I like to buy lawn seed from a company that names the ingredients in a mix, or that sells the varieties separately, instead of buying an "all-purpose" or a "shady-spot" mixture whose only recommendation is the honorable name of the company. There are fescues and fescues, and various grades and kinds of bluegrass. The one sure thing is that a bargain lawn seed is a poor bargain. After studying a half-dozen catalogues and nearly as many reference books, I gather that Merion Kentucky Bluegrass is generally accepted as the bluegrass of the highest quality, but the variety called Newport Kentucky Bluegrass gets started faster, so it is useful in the mix. Pennlawn Red Fescue, a grass developed at the Pennsylvania Agricultural Experiment Station, appears to top the fescues. So for our particular lawn I like the Harris M–61 Blend, which combines these three grasses. It may be obtained from the Joseph Harris Company. If I lived in the South, I would probably choose my seed from the varieties suggested in "Park's Flower Book," the catalogue of the George W. Park Seed Company, which has a good selection of grass seed and grass substitutes for warm climates. Park recommends *Zoysia japonica*, a Korean lawn grass that is increasingly used in the South; it is hardy as far north as southern Massachusetts, but where frosts come early it is unsightly, because it turns brown with the first touch of cold and greens up slowly in spring. (A brown lawn is also a fire hazard.) In the semitropical South, Bermuda grass and a grass called St. Augustine, which is very like crab grass, seem to be the most popular—at least in the part of Florida I know best. (If we Northerners think we are slaves to our lawns, we should remember the poor wretches of the semi-tropics, who must usually set out their lawn sprig by sprig instead of seeding it, and who must water it daily a good part of the year.) There are, too, substitutes for grass—ivy, myrtle, pachysandra, and other ground covers—but few of these can be said to provide the pleasures of green turf. The latest of the substitutes to be widely touted is dichondra, a creeping perennial with kidney-shaped leaves, which can be mowed. It is not winter-hardy in the North, but Park's catalogue recommends it for the warmer states. *Taylor's Encyclopedia* describes dichondra as a creeping tropical vine of the morning-glory family, "of no hort. interest except for *D. repens*, which is widely used in Calif. and other desert regions . . . where lawn grasses will not grow," but adds that "along the Gulf coast it is a pestilential weed," so one should think twice before sowing dichondra.

To return to Geoffrey Taylor and his "Note," Dr. Taylor points out that although every gardener is indebted to Mr. Budding and his machine, the only true labor-saving device for lawns would be a grass that did not grow tall enough to have to be cut. All through the nineteenth century, substitutes for grass were tried out in England—the mosses, the sand-worts, and, finally, *Spergula pilifera*, or spurry, which caused a flurry of excitement when it was introduced by an English nursery gardener in 1859. But spurry and the others failed to stay green in all kinds of weather, and the search for grass substitutes was abandoned, at least for the time being. (We are reviving this search, as I have implied, but thus far there is still no adequate substitute for turf.) "And so," concludes Geoffrey Taylor, "the no-trouble lawn had not arrived after all. And one must suppose that there is little use in looking for it; that this, like all other seekings after perfection—the quest of the Philosopher's Stone, or the Elixir of Life—also is a vain thing." Dr. Taylor should know, but I wonder. Nearly every garden catalogue of the year boasts a new flower that is shorter or tinier than any of its kind ever grown before: dwarf zinnias, stunted snapdragons, miniature roses, Lilliput this and pygmy that—many of them to me monstrosities in reverse. If the chemists and hybridists of modern horticulture can dwarf our noble flowers at the rate they are now doing and can breed midget corn, why can't they turn instead to dwarfing lawn grass, one of the few green things that really need to be dwarfed? I wouldn't put it past them, if there is enough demand, to come up eventually with a bluegrass two inches tall and a creeping fescue that grows no higher than a moss. There would be a fortune in it, too, if the figures in a recent *Saturday Evening Post* article are to be believed. The author, James A. Skardon, states that our twenty-five million home lawns cover five million acres of our lawn-happy land.

The development of the mowing machine has not stopped, though. A glimpse into the 1962 catalogue of the National Farm Equipment Company might astonish the inventor of the Moto-Mower, let alone Mr. Budding. It even astonished me. I don't pretend to know anything about machinery, and I cite National's catalogue only as an example, not as a recommendation. The title of this concern, by the way, seems to me a misnomer, unless I am out of date in thinking it odd that a farm-equipment catalogue should devote so much space to patio furniture, swimming pools, cascading fountains, billiard tables—and lawnmowers. At National you can buy just about any sort of mower, including five riding power mowers on which a man astride, after fighting his way home from the office in his car, may merely continue at the throttle. If he has bought the most sophisticated of the five—National's "25" Lawn-Rider, which has a push-button starter, forward and reverse shifts, and variable speeds—he will have spent "only $229.50." Having made this investment, he'll almost surely want such attachable extras as a fertilizer spreader ($24.75), a lawn aerator ($36.75), a trailer cart ($49.50), a snowplow ($29.95), and a Magic Fog, Jr., for broadcasting poisonous weed killers and insecticides (a mere $11.95). And if he *really* cares for status, he ought to pay out $67.50 more for a National Edger-Trimmer and $139.50 for a "Huffy Outdoor Vacuum Cleaner," which, with its "high velocity jet vacuum action" and "32-inch swath," its "giant self-discharging bag" and "semi-pneumatic, puncture-proof tires," will pick up just about everything, from grass clippings and leaves to barbecue refuse and Junior's toys. My own preference would be to retain our aging power mower and hand clippers and invest the same amount of money in the services of a high-school boy. Yet I do not begrudge the suburbanite his modern machinery. Seated on his Lawn-Rider, he must often drive in dream as he cuts his own "high mowing" or "upper forty." Very likely he can even sniff, in his mind's nose, a barn loft full of cured hay for his purebred herd.

I came late to the history of gardening, and have been trying to catch up on it in the past year. In addition to the Geoffrey Taylor books and the Hedrick *History*, I have read with profit Miles Hadfield's *Pioneers in Gardening* (Routledge & Kegan Paul, London, 1955), which covers too much to go very deep but at least takes what one might call a good lick at the whole development of the garden by discussing its pioneers and innovators in nearly every field. Starting with Theophrastus, the Greek "Father of Botany," Plinius Secundus, the Roman naturalist, and Disoscorides, whose *Greek Herbal* (or *Materia Medica*) was consulted and plagiarized for fifteen hundred years, Mr. Hadfield carries us by only a couple of leaps straight into Elizabethan England and to Thomas Hill, who wrote the first book on gardening to be published in English. From there the author marches onward through the centuries, pioneer by pioneer, to the present. It isn't, perhaps, a distinguished book, but it is a useful and orderly survey, and I believe it would be a good introduction to a far more thoughtful and better-written book, *Men and Gardens*, by Nan Fairbrother, published by Knopf in 1956, which I like so much that I want to speak of it now for those who may have missed it, as I did, when it was published. The title itself may explain why I value this book. Flowers one can like or even love for themselves, but gardens inevitably relate to Man—I use the general noun if only to include Woman, who fooled around the Garden of Eden and has rarely been persuaded to keep out of the flower beds since—and though the earliest gardens were grown for food and medicine, the garden very early became a philosophical as well as an aesthetic concept. Miss Fairbrother therefore takes us on a philosophical journey from the beginning gardens to today, omitting only the entire nineteenth century, whose contribution to the garden she considers insignificant. (On this I wholly disagree; the last century, hideous as were many of its gardens, was notable for its horticultural explorers and plant collectors, if for nothing else, and the stories of their strange, courageous, and often disastrous travels are to me as interesting and important as any part of horticultural history. The Hadfield and the Geoffrey Taylor books can fill this odd gap.) *Men and Gardens* is scholarly and very "literary"—

too full, perhaps, of apt, wise, and often delicious quotations from the many obscure or famous books that the author lists in her bibliography. What makes it stimulating is her thoughtful search into such themes as why men have gardened, how mores and history itself are reflected in garden style, just what a garden *is*, and what we want of our gardens today. Miss Fairbrother has an interesting mind and an adroit pen, and the journey her readers take through the Dark Ages and their monastic gardens, the Middle Ages and their romantic ones, the fantastic gardens of Tudor England, and the formal *jardins d'intelligence* of Le Nôtre and his followers, in both France and seventeenth- and eighteen-century England, is greatly enriched by having her as a guide.

A new book by Buckner Hollingsworth, *Her Garden Was Her Delight* (Macmillan), is based on a promising idea: to break new ground by writing about the little-known women gardeners, botanists, botanical artists, plant collectors, and garden writers who have played a part in the horticultural history of this country. The author's method is suggestive of Geoffrey Taylor's, but whereas Dr. Taylor covers only the Victorian era in England and devotes more space to each portrait (he wrote full-scale memoirs of just four people in his first book—the two Loudons, William Robinson, the garden revolutionist, and Reginald Farrer, the gifted and eccentric plant collector), Mrs. Hollingsworth covers the field from the early Colonial days to the very near past, and gives us twenty all too brief biographical sketches of women whose birth dates range from 1607 to 1863. The brevity is not the author's fault. Obviously, a lot of research has gone into this book, often with pleasant and surprising results, but when I had finished reading it I had the slightly uneasy feeling that in the early biographies there had been too little source material to flesh out the bare bones of fact, and that the American women gardeners who came later, about whom much more could be learned, were for the most part either not very interesting as human beings or not very important in the history of horticulture. In fact, the two non-Americans in the book—Jane Loudon and Gertrude Jekyll—are the only really tall figures in Mrs. Hollingsworth's landscape, and, probably because so much has already been written about them, she gives us only a short outline of their lives. Nevertheless, I enjoyed the book, and I think many dedicated women gardeners will want to read it, if only to learn a little about their dedicated predecessors—women like Alice Fenwick, who came with her husband from England to Connecticut in 1639 and helped him establish the first New England nursery of fruit trees in the then wilderness of Saybrook, just as he helped her "dress" the first American ornamental garden. One of the most interesting chapters is about Martha Logan, of Charleston, South Carolina. The title of the book refers to her and is not, as might be supposed, the work of the author herself, writing sentimentally, but a quotation from a letter written in 1761 by John Bartram, a Philadelphia botanist and plant collector, to Peter Collinson, a London botanist: "I received a lovely parcel [of seeds and cuttings] in the Spring from Mistress Logan, my facinated widow. . . . Her garden is her delight." Mistress Logan was a sedulous as well as a delighted gardener, and actually made her living, after her husband's death, by selling plants and seeds; she also published America's first book on gardening, *The Gardener's Kalender*. Little now is known about this true pioneer, but Mrs. Hollingsworth has gathered up the fragments of fact available and made the most of them. The chapter on Jane Colden, our first woman botanist, is also interesting, but again the facts are skimpy and too much has to be speculation. (Jane's father, Cadwallader Colden, appears to have been the giant of *this* family. Hedrick calls him New York's most distinguished botanist of the eighteenth century.) Then there was modest Maria Martin, who, with Audubon's encouragement, became a good botanical artist and supplied some, but by no means all, of the botanical backgrounds for *The Birds of America*. The final sketch is of Mrs. William Starr Dana, the author of *How to Know the Wild Flowers*, the popular book of wild-flower identification, published in 1893, on which so many of us grew up. I have a copy of it still—a 1906 edition—and find the book, because of its grouping of flowers by color, easier to consult in haste than the more modern and more scientific wild-flower books that group the flowers by their botanical families. If one isn't a botanist but an amateur like me, one can spend hours searching out a flower's family, not to speak of finding the flower itself within the family. I shall always be grateful to Mrs. Dana for her sensible and unscientific identification scheme, and I'm glad Mrs. Hollingsworth seems to feel the same way.

Excerpt from a Maine garden diary:

May 2, 1962

Just in from planting a Henry Leuthardt double-horizontal-cordon apple tree against the north brick wall of the terrace. Frost last night and bitter cold today, with an east wind off the sea so congealing that I longed for gloves. The Leuthardt tree is a Chenango Strawberry summer apple, a perfectly trained specimen and in perfect condition on arrival—the fulfillment of a dream. We also planted a Leuthardt dwarf Red Astrachan in the orchard, to keep us supplied with deep-dish apple pies when the old Astrachan tree gives up for good.*

When I called the double-cordon apple the fulfillment of a dream, I meant that I'd wanted for years to achieve an

* The Leuthardt tree is still going strong, though ineptly pruned. The Red Astrachan was eaten by the deer.

espaliered apple, but our attempts to train a dwarf McIntosh against this wall had been a ludicrous failure. The double cordon is by no means the most beautiful of the many espalier forms Henry Leuthardt has to offer, but it is handsome enough, and I chose it because it exactly fitted the north wall, which is low. The east wall would have been a better location, being higher and having a more favorable exposure, but the east wall serves as backstop for our woodpile from midwinter, when the wood is cut and fitted, to midsummer, when it is dry enough to store in the woodshed. This puts the pile of wood in the foreground of our view of the cove, the bay, and the mountain, but any northern country dweller knows there's no handsomer view than that of a year's supply of wood, well fitted and neatly stacked. By all means send to Mr. Leuthardt for his catalogue and handbook "How to Select, Plant, and Care for Dwarf Fruit Trees, Espalier Trained Fruit Trees, Hybrid Grape Vines." You will find excellent photographs of the other lovely espalier forms, such as the four-armed, six armed, and eight-armed Verriers and the double- and triple-U forms, all available in both apple and pear, and the fan shape, available in peach, apple, pear, and nectarine. Or you may prefer to grow a lattice Belgian fence; five espaliered trees in a row will make a diamond-lattice effect of apples or pears. The handbook explains how trees are dwarfed and trained, why grafting on Malling rootstocks makes the Leuthardt trees especially desirable, and what Malling rootstocks are. I found this booklet a model of clarity and foolproof instruction—an exception to the rule that reference books are more helpful than catalogues. There isn't a question on cultivation, pruning, or general care that it fails to answer. Mr. Leuthardt explains in his introduction and in his letter to customers that he has devoted fifty years to growing dwarf fruits, the last thirty of them in this country, and that he comes of a long line of Swiss who have excelled in the propagation of fruit trees. He never boasts, though, and his publication inspires confidence. It lists a most interesting collection of dwarf fruit trees. Many are European, English, or Japanese varieties, but he offers only such apples, pears, plums, peaches, apricots, nectarines, cherries, and quinces as he has learned by experience will survive our rugged climate. Some of his apples are old favorites from the American past, some are the best of the modern apples; some of the pears, too, are the old familiars, but many are choice French varieties; the plums are mostly Japanese. When I read that one of the apricots was a Moorpark, I could hardly believe my eyes. Moorpark? Straight out of *Mansfield Park*! I looked it up to be sure my memory was not playing me tricks. Sure enough, a "moor park" (as the author spelled it) was the very apricot Fanny Price's widowed Aunt Norris boasted of having planted at the parsonage the spring before her husband died. You will remember that the new incumbent, Dr. Grant, who loved his food, took her down at once by saying that she had been imposed upon—that the parsonage apricot was an inferior variety, with as little flavor as a potato. Mrs. Norris's retort was characteristic: "Sir, it is a moor park, we bought it as a moor park, and it cost us—that is, it was a present from Sir Thomas, but I saw the bill, and I know it cost seven shillings, and was charged as a moor park." A one-year-old dwarf Moorpark apricot from Leuthardt costs $2.50—not much of an increase over Sir Thomas's seven shillings in the century and a half since *Mansfield Park* was written. Although apricots are not among my favorite fruits, I am tempted to plant one in memory of that lovely Jane Austen squabble.

Mr. Leuthardt says that all his fruit trees can be planted in spring or in fall. His catalogue also lists grapes, including my favorite Delaware, which one seldom sees in the markets nowadays, and also a modest selection of nut trees, currants, raspberries, gooseberries, and blueberries. No strawberries.

For strawberry plants, and for just about every other kind of berry, the catalogue I have found to be the most comprehensive and satisfactory is Rayner's "Berry Book," put out by Rayner Brothers, Salisbury, Maryland. The raspberry and blackberry bushes we bought from Rayner last year and this, as well as some Rayner asparagus roots, were exceptionally fine on arrival and have performed well for us. Considering the "Berry Book" just as a catalogue, I admire it for its careful cultural advice, its photographs, in both color and black-and-white, and its authoritative tone. Rayner, like Leuthardt, is helpful to the home grower. There is a page comparing the characteristics of the twenty-six strawberry varieties offered and telling how to choose the ones that are best for each region; it also groups the varieties the firm considers the best for flavor, for quantity, for size, for freezing, and for preserving. Every strawberry listed has been grown from virus-free foundation stock furnished by the Department of Agriculture or the State of Maryland.

Blueberries are another Rayner specialty. They can be grown from Maine to North Carolina, from Michigan to Missouri, but if you live in Maine, where the wild blueberry is one of the main crops, it just doesn't occur to you to plant the large cultivated blueberry, which has less flavor. We used to keep up our wild-blueberry piece, but it has petered out because we no longer burn it, cut down the alders, and do all the rest of the work needed to insure a plentiful supply of berries. Every market in our region sells wild berries in season, and the stores conscientiously try to have them picked by hand from unsprayed bushes. But in the cities one can't be sure. Some berries may have been raked in commercial blueberry barrens where the bushes are heavily dusted or sprayed with chemicals to keep down the blueberry maggot and other pests. Almost all commercially canned blueberries come from sprayed stock. I'd rather eat a maggoty blueberry any day than a sprayed blueberry, so a few home-grown cultivated bushes, planted conveniently near the house, may be the answer, especially since Rayner lists one variety that is "from a wild type," the Rubel. It won't be the same, though, for it is a huckleberry, and every resident of Maine knows that the huckleberry can't compare with the wild low-bush blueberry. All the berries I have discussed should be planted in spring, but because the preparation of a good strawberry bed or a raspberry or blackberry patch takes time and planning, it would be well to make ready for a spring plantation before winter sets in.

Another specialty catalogue that interests me is titled "Bamboo for Cold and Warm Climates." It is put out, I regret to say, by a nurseryman who prefers to call himself "The Bamboo Man" instead of using his own name, James J. Coghlan. Bamboo, as I've seen it in Florida, appeals to me because of its dense, grovelike growth, its feathery leaves, its straight-as-a-pencil stems, and its great attraction for birds, which find it excellent cover. I had assumed it could be grown only in the South, and this is so of the true bamboos, the *Bambusas.* However, in his catalogue, which is heavily Japanese in flavor, Mr. Coghlan lists several cousins of the bamboo that he claims can survive temperatures as low as twenty-three degrees below zero, and have survived them in his nursery. Whether the Bamboo Man's bamboos are as hardy as he believes I have no way of knowing without trying his stock, which I have not yet had a chance to do. I don't doubt his word, but I know there are many plants that, even though they will live through a severe cold spell, provided it does not last long, are not actually hardy in zones where sub-zero temperatures may last for days or weeks. *Taylor's Encyclopedia* says that *Phyllostachys*, two varieties of which the catalogue recommends as *very* hardy, will grow only from Zone 5 southward. However, a root division of *Phyllostachys aurea*, often called the fishpole bamboo, costs only $3, and might be worth a try, even in chilly Zone 3, where I live. I can't imagine anything much more fun than to grow one's own fishpoles and garden stakes. Bamboo is a grass, not a tree or shrub, and is not subject to insect pests; it can be shipped and planted at any time of the year, and it has any number of uses—as a food (recipes are given), as winter forage for animals, as lumber, as ground cover (a dwarf variety of bamboo that can be mowed). The tall sorts are ideal, too, for chicken runs, windbreaks, hedges, and canes. Bamboo is also, once the roots have taken a real hold, the fastest-growing plant known. The new shoots emerge at the maximum diameter they will attain, and after they are three years old, if we are to believe the Bamboo Man, they literally leap, growing as much as three feet a day. And there are dwarf bamboos and other varieties good for bonsai, the Japanese art of training plants and dwarfing trees. The catalogue tells how to make bamboo into bonsai plants, and it includes bonsai materials, such as pottery containers imported from Japan. It lists, as well, a number of unusual Oriental trees and shrubs, and even a selection of Japanese stone lanterns for those who want to make their gardens *echt* Nipponese.

Before we leave the Orient, I must mention the new book by Ernesta Drinker Ballard, *The Art of Training Plants* (Harper), which is perfectly described by the subtitle on its jacket: "Extending the Oriental Art of Bonsai to Western Conditions and with Decorative Plants." Mrs. Ballard is the author of *Garden in Your House*, my house-plant Bible, so I read the new book eagerly, even though I had long since decided that life was too short to get involved in the new

Occidental fad of growing dwarf trees in the ancient Oriental manner. I was under the impression that it took decades for bonsai trees to become interesting, and that many of them were at their best only after they were centuries old. This turns out to be true, if one wants a really classic gnarled tree, but Mrs. Ballard's book opens new vistas, for she not only explains this ancient art of gradening in miniature but shows how one may train native American trees that have been stunted naturally and twisted into interesting shapes by the elements, and how, by using some of the bonsai tricks, one can make ordinary house plants far more pleasing to the eye, whether they are herbaceous or woody plants, succulents, bromeliads, or epiphytes. The book has excellent photographic illustrations, and one, of a decoratively trained slipper gloxinia, has inspired me to try to make my big pure-white gloxinia, which last year bore more than fifty blossoms, a trifle less sprawling and undisciplined. Like all Mrs. Ballard's books, this one is full of common sense and of clear advice, based on her own experience, and it should be a useful adjunct to *Garden in Your House*. For those who seriously want to learn to make tiny dish or tray gardens—the word "bonsai," by the way, is composed of the Japanese characters meaning "tray or pot" and "to plant"—all the instructions for dwarfing by root pruning and other care are given, and an appendix lists sources of bonsai material and plants.

I have at hand the new catalogues for three of the nurseries she lists as having used. One is White Flower Farm; its "Plant Book" has a good short list of trees for what it calls "Big Bonsai," and Italian pottery "ring pots" in which to grow them. These trees would not be the miniature *Mame* bonsai trees but merely manageably small potted trees for patios. Mayfair Nurseries (Nichols, New York) has a longer list of dwarf conifers and shrubs, many of them of the proper size for miniature gardens and bonsai work. One of Mayfair's specialties is heather and heath plants. The catalogue of Girard Nurseries, at Geneva, Ohio, is an attractive booklet, illustrated with color and black-and-white photographs, which includes many full-size evergreens, azaleas, shrubs, and deciduous trees, but its first four pages are devoted to bonsai trees and bonsai pots and dishes—reproductions of Japanese shapes in inexpensive but good-looking glazed American pottery. Group I of the trees includes baby ones suitable for bonsai culture, two to three years old, in two- to three-inch pots, and they are an interesting collection; Group II includes well-established dwarf trees, many of them unusual, grafted on four-year seedling understock, and already growing in bonsai pottery containers; Group III includes four- to five-year-old plants, which have been root-pruned for dwarfing several times and are already planted in bonsai pottery. A dozen varieties of this last group are also available in eight- to ten-year-old plants, growing in eight- to ten-inch bonsai containers. Some of the Girard trees are so rare and sound so easy to handle, considering the good start this nursery gives on the root-pruning, that I am tempted to experiment with a flowering bonsai dogwood, or a three-stemmed ginkgo, or a contorted Hinoki cypress, but I shall resist, because of my plethora of house plants and because, too, these little trees in shallow pots cannot live if they are not faithfully watered once or twice a day. As Mrs. Ballard points out, the omission of water for only a single day may prove fatal. I am faithful to my house plants in my fashion, but not as faithful as all that.

The American Medical Association's lobby is often cited as the richest, most powerful lobby in Washington, but to judge by my mail in the past year, David Burpee's one-man lobby to have the marigold named as our national flower must run it a close second. Indeed, there was a period around the turn of the year when a marigold message seemed to arrive in every mail from Mr. Burpee himself, from his public-relations firm, or from the W. Atlee Burpee Company, of which he is the president. Burpee is our largest general seed company, and one of its specialties is hybrid marigolds. Presumably it sells more marigold seed than any other seed house in the world; certainly it works harder at it than any other. That flood of marigold publicity started with a Christmas greeting from Mr. Burpee, enclosing several packets of marigold seed, and was followed in quick succession by a sheaf of glossy prints showing the new varieties, a landscape gardener's plan of a marigold garden, more sample packets of seed, more glossy prints, and, finally, a publicity release from a Washington firm called Earle Palmer Brown & Associates. As I have said before in these pages, I like marigolds, I grow marigolds, and I defend them against those who find the sharp herbal odor of their foliage offensive. I happen to enjoy this spicy aroma, especially in the modified form arrived at by the Burpee hybridists. I was also pleased as well as entertained, a couple of years back, when Mr. Burpee in person went to Washington, bearing marigold boutonnières for the legislators, and, at a Congressional hearing on naming a national flower, faced down the Senatorial proponents of grass, the corn tassel, the rose, and the carnation; I agreed with him that none of these four nominations was suitable. The rose is the national flower of England and seven other nations; the corn tassel is not a true flower but only one part of the corn's blossom; the carnation, as the word is used in this country, connotes chiefly the florists' hothouse flower, which is a hybrid of the garden pink, or dianthus, a plant native mostly to the northern sections of the Old World, including Russia; as for grass, its flower is inconspicuous and, to quote Mr. Burpee, "Who wants a national flower you can walk on?" But to my mind the arguments against the marigold are equally cogent. It is an annual, not a perennial with the staying power a national symbol should have; many people dislike the scent of its leaves; the chief garden varieties are commonly, if erroneously, known as French marigolds and African marigolds. Actually, both are Mexican. However, I did not worry unduly about our transmuting an Aztec flower into an American one until I received that publicity release. It started off with nothing more alarming than these press-agent headlines:

AMERICAN MARIGOLDS

FOR NATIONAL FLOWER

LOBBYIST USES MARIGOLD PETALS

OF INFLUENCE TO PRESSURE CONGRESS

Reading on, though, I was startled by the announcement that bills nominating the "American" marigold had been introduced by Senator Everett Dirksen (R., Ill.) and Representative Willard S. Curtin (R., Pa.). Lately, the Congress has been engaged in more urgent matters, but one can never tell when the silly season will set in on Capitol Hill or when our legislators will want to take a breather from the Cold War. I hurried to my faithful reference books, *Hortus Second* and *Taylor's Encyclopedia*, to be sure of my facts in controverting the claims made by Mr. Burpee's Washington "Associates." The release, a folksy document, is composed mostly of purported quotes from the Marigold Master, or, as it describes him, "the gardener's mail-order green thumb." Well, in the supposed words of Mr. Green Thumb, who surely knows better, "American marigolds is [sic] the most appropriate candidate to represent the United States, botanically." Yet both Liberty Hyde Bailey and Norman Taylor say that all four species of *Tagetes*, the plant we have hybridized into garden marigolds, originated in Mexico. "American marigolds," continues the Green Thumb, "are native to the American Continent only." But *two* continents are involved, for Liberty Hyde Bailey states that *Tagetes* is now to be found growing wild from New Mexico to Argentina. The release winds up with a bit of homely advice singularly inappropriate to so confused a document: " 'It helps clear a man's mind to get his hands in the soil,' advises the sixty-eight-year-old seedman."

Feeling as I do about the marigold as the flower of the United States, I've decided to start a flower lobby of my own. I thought of nominating the moccasin flower (or lady's-slipper), our handsomest native orchid—it would be a nice tribute to our earliest inhabitants—but the moccasin flower does not grow in all our states. May I therefore suggest the humble but glorious goldenrod, which grows wild in every state of the Union and spreads its cheerful gold in August and September from Maine to the Pacific, from Canada to Florida and New Mexico? It is as sturdy and as various as our population; there is delicate dwarf goldenrod, silver goldenrod, tall yellow goldenrod in a multitude of forms and shapes—spikes, plumes, and panicles of native gold. *Hortus Second* lists fifty-five species, only *one* of which is not native to the United States, and one or another of these fifty-four grows in every sort of soil. Descend into a bog and there, growing wild, is goldenrod; climb a mountain and there, between the crevices of boulders, is goldenrod; follow the shore of the sea and goldenrod gleams along the edge of the sands; drive along our highways from coast to coast in August and September and the fields and ditches are bright with goldenrod, unless the state you are driving through has destroyed them with chemical sprays. The very ubiquity of the flower has given it a bad name as an irritant to hay-fever victims, but I've recently read that it is the ragweed and flowering grasses growing alongside goldenrod that are the villains during the late-summer hay-fever season. The goldenrod also has the great advantage—if it were to be our national flower—of owing nothing to man, of enriching no seed company, or companies, and of being as wild as our national bird, the eagle. Canaries, like marigolds, presumably thrive in all fifty states, yet no one would dream of nominating the canary as the national bird.

Who, for lack of a garden, hasn't gathered armfuls of goldenrod to bring cheer into a bleak summer cottage or camp? Who, even *with* a garden, hasn't put a jar of goldenrod or asters on the porch in August? Thus, we think of it as common, yet goldenrod, whose Latin name is *Solidago*, is not a humble flower. It must have been taken to England from North America by one of the early British plant explorers, for American goldenrod has long held a cherished place in many an English herbaceous border. Now, ironically, it is being exported back to us and offered in the nursery catalogues as something very special. You may this year buy, at $1.50 a plant, from Wayside Gardens a hybrid variety called Solidago Primrose Cascade. The Wayside catalogue describes it as "a must." This may be a hybrid of the one European species, but I doubt it, for American wild flowers, sent to England in the nineteenth century and earlier, became a craze with the most opulent of the Victorian flower gardeners.

Goldenrod has many unusual uses. It can be made into a crown of gold for a Harvest Queen or a necklace for a child. Sewn onto a white bedsheet, it makes excellent cloth of gold. I lay under just such a sumptuous covering one summer day in my early teens when there was a regatta on Lake Chocorua. This was in that almost forgotten era before children were sent off to camp, when summer pastimes for the young were innocently contrived at home. On that bright New Hampshire mountain day so long ago, a parade of decorated canoes and rowboats was the first event in an afternoon of water sports, and, like every other family on the lake, we wanted to be in the parade. My sisters and I began the day by gathering bucketfuls of goldenrod, which my aunt spent the morning stitching onto a sheet in an intricate pattern—no stems or leaves showing, only the plumy golden blooms.

Alfred, Lord Tennyson, must again intrude upon this report, for he was responsible for our flowery cloth of gold and, in part, for my regatta-day ordeal. His intrusion is not wholly inappropriate to a discussion of flowers, since Geoffrey Taylor points out that although Tennyson's Queen had little interest in flowers, her Poet Laureate was the most garden-conscious of all the Victorian poets. Perhaps this was why he was so much read by my nineteenth-century garden-loving elders and why it was entirely natural for the family to decide to make our canoe represent the death barge that carried "Elaine the fair, Elaine the loveable,/Elaine, the lily maid of Astolat" down the river to Camelot, bearing in her white hand her last message to Lancelot, who, to his shame, at the moment loved only "the wild Queen" Guinevere. So while Aunt Caroline sewed, the rest of us were busy all morning draping our green Morris canoe in black and devising costumes for Elaine and the dumb oarsman. I, only because I was the smallest and weighed the least and had the fairest hair, was to be Elaine, dressed in white cheesecloth, and my father, cloaked in black, the spellbound oarsman. (I remember swathing a peaked Mexican straw hat with my big black silk middy-blouse tie, to make a suitably lugubrious covering for his bald head.) We had taken a heavy wooden bathhouse door off its hinges and laid it on top of the gunwales of the canoe, leaving open only the stern seat. So, after my sisters had covered my summer-tanned face and arms with a heavy coating of white talcum powder, to achieve the pallor of a defunct lily maid, I was gingerly laid out on the door, and my long hair, which

I usually wore in pigtails, was spread out under my shoulders. My arms were crossed on my breast, the letter to Sir Lancelot was placed in my right hand, and the goldenrod cloth of gold was drawn up over me and my bathhouse-door pallet. Then it became my father's terrible task to seat himself in the stern and paddle this top-heavy, teetery craft and his youngest daughter a mile or more, from the head of the big lake, where we lived, through the narrows to the small lake, where the regatta was assembling. The sun beat down upon us, and my powdered face and nose grew redder and redder. My father couldn't stand the weight and heat of his hat, so he laid it at my feet. A light breeze sprang up, and the overladen canoe became even harder to handle. His usually cheerful round face was soon as "haggard" as the oarsman's of the poem, and he constantly cautioned me not to move a muscle. I didn't; I was scared stiff. As we drew in sight of the judges' float, my father resumed his peaked hat, and I, closing my eyes, put on what I hoped was a demurely love-lorn expression. Then we circled endlessly in the parade. At one point, the oarsman removed the letter from my tanned fingers (all the powder by now had melted away), paddled up to the float, and handed it to one of the judges, who read it aloud to his colleagues, just as King Arthur read out Elaine's message to his assembled knights:

> "*Most noble lord, Sir Lancelot of the Lake,*
> *I, sometimes call'd the maid of Astolat,*
> *Come, for you left me taking no farewell,*
> *Hither, to take my last farewell of you.*
> *I loved you, and my love had no return,*
> *And therefore my true love has been my death.*"

Without this message, I fear we would not have been recognized and would not have won first prize. For years, I kept the trophy—a framed photograph of Mount Chocorua with the lake at its foot—knowing all along that it was my aunt's lovingly sewn goldenrod cloth of gold that had won it for us.

My seed order to Burpee this year, in spite of those gift packets, included a packet of Climax Marigolds in mixed colors. These are the current Burpee F_1 hybrids of the tall Aztec marigold strain, and I wouldn't be without them. In September and October, their tall double blossoms are a part of our autumn scene, outdoors and in, for I often put a big copper jar of them in the living room just before the killing frosts come on. The three brand-new 1962 Burpee marigolds are Orange Glow, with quilled sunflowerlike blossoms; Penn State, with gold flowers that imitate chrysanthemums; and Primrose Climax, a soft primrose-yellow hybrid in the Climax series, and a real beauty. There are dozens of other sorts, too, filling six pages of the catalogue—the Miracles, the White Golds, the Carnation-Flowered, the Chrysanthemum-Flowered, the Sun Giants, the French Dwarf Singles, the Dwarf Doubles, and the Pygmy Doubles. I love them all—but please, ladies and gentlemen of the Congress, *not* for our national flower!

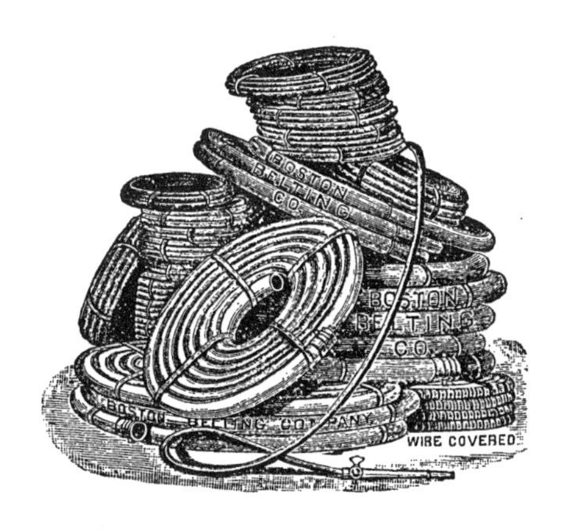

8

An Idea Which We Have Called Nature

Dutch garden with an arbored alley and beehives.
By Jacob Cats, 1622

December 18, 1965

We are surrounded with things which we have not made and which have a life and structure different from our own: trees, flowers, grasses, rivers, hills, clouds. For centuries they have inspired us with curiosity and awe. They have been objects of delight. We have recreated them in our imaginations to reflect our moods. And we have come to think of them as contributing to an idea which we have called nature.—*Sir Kenneth Clark*

How we are re-creating in our imaginations Sir Kenneth's list of our objects of delight in this age of modern architecture, crowded cities, and sprawling suburbs is the theme of *Modern Gardens and the Landscape*, by Elizabeth B. Kassler, who introduces her book with Sir Kenneth's words. This slim but rich volume of ninety-six pages is a publication of New York's Museum of Modern Art (it is distributed by Doubleday) and consists chiefly of excellent photographs, most of them black-and-white, five in color, each described and commented on in some detail by Mrs. Kassler. It is, however, her preface that I value the most, for it contains some of the wisest thoughts I have read on the philosophy of landscaping and gardening and on how the aesthetics and pressures of our time have affected the modern landscape artist. After giving us our relationship with the past—our direct inheritance from Europe and England, the influence of the ancient gardens of China and Japan (where rocks were sculpture and plants were trained into animate sculpture), the Moslem contribution of water (flowing, leaping water for action, placid pools for reflection, or water cunningly contrived to irrigate arid land) —she concludes that the contemporary designer must create in terms of the present. Planting "must appear to be of its place, not on its place," and she warns the artist that he must, when he takes hold of earth, water, and plants, beware lest he "destroy through his act of possession the genius of that which he has sought to possess."

Roberto Burle Marx, of Brazil, is the leading modern landscape architect of our hemisphere, and a number of the photographs in Mrs. Kassler's book illustrate his skill at combining native plants with modern buildings in South America. A fuller sample of his style is given in a large picture book (too big to read in bed), *The Tropical Gardens of Burle Marx*, by P. M. Bardi (Reinhold), with photographs in color and black-and-white. Mr. Marx's style is varied to suit the locale, and his favorite, when he has space enough, seems to be compositions of "well thought out masses: now a contrast volume, now an isolated complex, marginal arabesques, and always strict control of the color scheme." His two most frequent devices are undulating curves of contrasting mass plantings—for example, wide swales of white plants, magenta plants, light-green plants, and grass-green plants in adjoining strips that swirl like the contour rows of our modern farms—and mass plantings in beds of free-form modernistic shapes that form an abstract design. Seldom is there a mixture of color in any one bed. It is all very effective and very, very up to date, yet an odd thought struck me as I looked at the illustrations. Had we come full circle, to a modern version of the now despised Victorian gardens, where plants were "bedded up" or "carpet-bedded" to make patterns on a lawn? The Marxian formality, with its massing of plants in curves of contrasting colors to make a pattern, is a sort of carpet or tapestry work of another kind, and it can sometimes bore me as much as the naïve Victorian formal circles and triangles of bright-colored plants set in little beds on green turf, even though, with Marx, the beds are asymmetric and the swales curve in fluid lines. I even wondered whether, if I were a patient in the Larragoiti Hospital in Rio de Janeiro, looking out my windows at Marx's beautiful modern garden below me, I might not tire of the formality of its circular cream-color cement space, with blue tile wall and pure-white cobblestone curving paths, and its masses of magenta, gold, and green plants. Wouldn't I prefer to look down, as I have during a long hospital stay, on a haphazard, unpatterned old-fashioned shrubbery and a few formal beds of roses or of spring bulbs, to be followed by summer annuals among which, one by one, the shrubs or flowers would come into blossom and offer me a new surprise for every week of a long convalescence? The very thought is lèse-majesté in the aesthetic code of today, yet I have to admit I had it. Too much fixed form in a garden, however abstract and imaginative, can be tedious, and to me it doesn't always contribute to "an idea which we have called nature." Many of the Marx mass plantings are colored by their foliage, not their flowers, though some have blossoms. What happens, I also wonder, to Mr. Marx's Picassian tapestries when some of the bright-colored flowers go to seed? Perhaps they are uprooted and replaced? Ah, well, labor is cheaper in Rio than it is in New York.

In spite of these doubts, I greatly admire Burle Marx for his originality and for the ferment he has started in the world of landscape architecture. I admire him, too, because he is a horticulturist and plant explorer, a gardener and botanist, as well as a designer of gardens and landscapes. He has sought out in the tropical forests of his continent trees of unusual form, studied their habits, and transplanted them to where they will thrive and become plastic forms that tie in with the buildings of Niemeyer and other modern architects. And Marx's small, intimate gardens can be just that—lush, private tropical retreats, with lotus in pools, or small made ponds crossed by stepping stones in the Japanese manner, surrounded by a fascinating variety of tropical trees and plants.

No carpeting here, no intrusion on the genius of the place. Professor Bardi, the director of the Museum of Art in São Paulo, has done justice in his text to Mr. Marx's versatility, for this astonishing man is not merely an artist using growing things as his medium; he is, as well, an abstract painter, an architect, a stage designer, a jewelry designer, a planner of cities, parks, and zoological gardens—a mover and doer all over the world. The Bardi text is in English in this edition; the picture captions are also given in German and Italian. The book unfortunately lacks both an index and a table of contents, so a photograph or reference can get badly lost if a reader wishes to study it again.

If Burle Marx appears, in the color plates of this volume, to use more magenta, red, and purple than I care to see en masse, these sometimes grating colors could be a mere matter of poor color reproduction. I am not well enough acquainted with the tropical vegetation of Brazil and Venezuela to know how true to life the colors are, and none of my North American reference books can help me out. Reproduction of color photography is always tricky, as is shown by another huge, deluxe volume brought out in the past year—*The Odyssey Book of American Wildflowers* (Odyssey) with photographs by Farrell Grehan and text by H. W. Rickett, who is the Senior Botanist of the New York Botanical Garden. The text is exemplary, but the color plates are an instance of how things can sometimes go wrong in color printing. Many of the three hundred color plates are breathtakingly lovely, yet a reader trying to identify a wild flower could be sadly misled. Mountain laurel comes out a baby blue instead of the white or pale pink it is; bunchberry has acid-green leaves instead of the good grass green of nature; the pale pink-blue-white of the delicate bluet is almost aquamarine; the violet blue of the wild blue flag is falsely shown as a harsh royal blue. These few examples of many distortions are clearly faults of the color printing, but the photographer, whose efforts to make his pictures works of art and not merely literal are praiseworthy, can also be misleading. I do not know how he achieved his lighting effects; some of them are very queer. The shy wild columbine of the East (*Aquilegia canadensis*), with its modest red-and-yellow flower, which grows in wooded rocky areas, is photographed in a blaze of orange red instead of against the green or gray or brown forest floor of its shaded natural habitat, and the wild larkspur of Oregon and Washington appears to be blooming on the edge of a forest fire.

I finally found Mr. Grehan's backgrounds and modern photographic effects so distracting that I turned for relief to the clear, pleasant, and quite literal photographs, taken for the State of New York in the early part of the century, that illustrate Homer D. House's famous book *Wild Flowers*, which is still the classic reference book for the wild flowers of our region, even though it was first published in 1918. In 1934, Macmillan brought out a popular one-volume edition, and in 1961 it reissued this useful, rewarding book, with new plates made from the original illustrations in the State Museum in Albany. All the flowers are shown not in natural settings but against a pale-gray background. Even so, how charming and true to life they seem! Whatever the recent progress in photography and color reproduction, it counts for little if it doesn't work.

Sometimes it *does* work. Almost any book issued by the Sierra Club proves this. Many by now must be familiar with the series of luxurious volumes, in what the jackets call "exhibit format," ten and a quarter by thirteen and a half inches, that is published by the Club and sold without profit. The books are expensive at that; the two I shall discuss are twenty-five dollars apiece. Most of the Sierra publications are dedicated to photographing nature—particularly the wilderness—in color, as exactly and beautifully as possible. I have not seen all twelve of the volumes issued thus far, which represent many different authors and photographers and many locales, from New England to the Pacific. Of the two I now have before me, which I bought because they deal with land I know, my favorite is *In Wildness Is the Preservation of the World*, with text by Henry David Thoreau, a preface by Joseph Wood Krutch, and photographs by Eliot Porter, who also made the excellent selection of quotations from the twenty volumes of Thoreau's work. Mr. Porter and Thoreau seem to be particularly attuned to each other; every short selection is apt to its accompanying picture, and vice versa, and each selection is a perfect thing in itself, whether descriptive, philosophical, satiric, or funny. Many of the photographs have the quality of a classic representational painting, something one could hang on the wall and live with in comfort—feathery elms in a snowy meadow against a gray winter sky, with weeds and low shrubs thrusting their brown stalks through the deep snow of the foreground, accompanied by Thoreau's wonderfully witty comparison of elms and politicians; a closeup of pond lilies photographed just as Thoreau described the ones he found in a muddy inlet of a meadow in 1853, perfect and gleaming white-and-gold, before the insects had discovered and invaded the blossoms ("It is remarkable," wrote Thoreau, that "out of that fertile slime springs this spotless purity"); barn swallows in their nest being fed by the mother bird, with Thoreau's observations on the musical-chairs process of swallow feeding by which each baby shifts in turn into the position to receive the next insect or is chastened by the mother if it tries to cheat. Only a few of the photographs lack the verisimilitude of nature; even with all the painstaking care in color printing and this photographer's skill, some of the greens—always difficult to reproduce—are a trifle off.

My second Sierra Club volume is *Gentle Wilderness:*

The Sierra Nevada, with a condensed version of John Muir's *My First Summer in the Sierra* as text, and photographs by Richard Kauffman, who, like Porter, is an extraordinary cameraman. He has ranged the Sierras since he was a child and knows how to photograph their bold crags and azure lakes, their towering or twisted trees, and their delicate flowers. He includes several placid snow pictures but seems determined to emphasize the gentler qualities of this rugged range. Perhaps the Sierras *are* the gentlest of our higher ranges, though those mountains are not so gentle, after all, as the survivors of the Donner party could testify. I longed for a Sierra blizzard. The greens are better in the Kauffman volume, which is No. 9 in the series; Porter's is No. 4. Thus, painstaking care in the color reproduction is paying off. Still, Kauffman, for all his virtues (new techniques, new photographic and printing processes), seems to me more of a literal photographer and less of an artist than Porter. Some of his pictures, however, do what hardly any painters can in showing bold, miles-wide sweeps of rugged landscape. It could be that my preference for the Porter photographs is purely a matter of personal prejudice, just because I know better and love more the quieter natural world of New England.

Another hefty, enormous book illustrated by photographs is the already well-known *Great Gardens*, by Peter Coats, with an introduction by Harold Nicolson, which was published in 1963 in England, and in this country by Putnam, which has changed the title to *Great Gardens of the Western World*, a title that describes more accurately the limits of Mr. Coats's thirty-six famous gardens of Europe, the British Isles, and America. The photographs, the work of many, are spectacular. I especially like the black-and-whites, though with a few exceptions the numerous color photographs—at least in the English edition, which is the one I have studied—are true in color and expert in photography. The text, descriptive and historical, is interesting, but I did feel that Mr. Coats was more at home in England and Europe—even in the Peterhof Garden, in Leningrad, his only Russian example—than in America. It is a peculiarly insular and English volume, all in all, often to its advantage. The author, though, has an unfortunate tendency to use superlatives in his chapter subheadings. The courtyards of the gardens of the Alhambra "form the oldest gardens in the world," a dubious statement if ever there was one; Chatsworth is England's "finest country house," and Versailles is the "greatest garden in the world." Well, perhaps, but some of these superlatives depend on an individual's taste in gardens. These are Mr. Coats's own very British enthusiasms, which is quite all right except when he exaggerates. Anyway, the book is a luxurious illustrated tour through France, Great Britain, Ireland, Holland, Portugal, Germany, Italy, Spain, Russia, and the United States, and the text reads like the spiel of a loving and knowing guide. Here you can see the panorama of Western taste over the centuries, from the most formal gardens of statuary, clipped trees, and architecture to the more informal landscape gardens of England and of Charleston, South Carolina.

Photographs, whether in color or not, are useful for illustrating broad-scale scenes of large gardens and landscapes, but the question remains whether the camera can ever really convey the essence and true beauty of an individual flower. I myself don't think so. There is something extra that the best of the flower painters add—the artist's subjective touch that makes his picture a work of art and yet shows its detail even more accurately than photographic color plates have thus far been able to. Compare the color plates of Pierre Joseph Redouté's famous paintings of the ancient roses with the hundred and twenty-eight pages of photographs in Bertram Park's *The World of Roses*, a big picture book published in 1962 by

Dutton. Mr. Park's color plates were made in England and almost all of them are quite good, at least for the roses I know and have grown. Only a very few seem to misfire. To my eye, his Michèle Meilland, my favorite light-pink rose, has the yellow-pink tones of Confidence rather than the pure, pale china pink of Michèle; his Sutter's Gold fails to catch the character of this fragrant, tawny-yellow, and slightly undisciplined rose; his Charles Mallerin almost but not quite gets the black heart of this darkest of red roses; his two-toned Duet isn't half as lovely as the real one on my terrace, and so on. Part of this may be a matter of climate and soils, since rose colors vary in different climatic conditions and soils. Mr. Park's is a useful book, however, showing a wide variety of roses, new and old-fashioned, and I recommend it as a guide to anyone who is just starting a rose garden. You can see what you'll be getting far better than in most rose catalogues. It is not, however, a book I want to go back and study again and again as I do to the various Redouté folios. The pictures are sometimes quite lovely, but the total effect is loud and the photographs never convey the total personality of a rose in the way an artist can. Lay the Park volume open alongside a Redouté volume and the comparison is devastating.

Happily, Graham Stuart Thomas's latest book, *Shrub Roses of Today* (St. Martin's), shows that the art of painting the rose is not entirely lost. The book contains eight water colors of moderate size and eight pencil drawings by the author, as well as photographs. Mr. Thomas is the acknowledged authority on old-fashioned roses and shrub roses, and has done much to revive an interest in growing these more natural, subtler, and less showy roses in an age when the modern hybrid tea rose engrosses most gardeners. He is a man of many parts. Gardens Adviser to the National Trust in Great Britain, he also has in his care, at Sunningdale Nurseries, of which he is the manager, probably the most nearly complete collection that exists of shrub roses, both old and new. Added to all this are his gifts as artist and writer. *Shrub Roses of Today* is a pleasantly written and informative book. It is a sequel to his *Old Shrub Roses* and takes up where that one left off in the history of shrub roses. The new book has a chapter on wild American roses and includes American hybrid shrub roses in its descriptions and lists; it is the best history there is of the development of modern shrub roses from their early forebears. I relish Mr. Thomas's complaints about the increasing garishness of color in the new hybrid teas and his objection to the blatancy of the scarlet, orange, flame, and yellow-toned reds that are now so fashionable. As he points out, these colors are not natural in roses and fit badly into any garden scheme because, being at the opposite end of the spectrum from green, they look harsh against the green background of every garden. He also has a fine chapter on fragrance and a practical one on the pruning and cultivation of roses as shrubs. As a floral artist, he is no Prévost or Redouté, but at least he does not imitate them, as so many more pretentious modern artists do when they paint roses, and his water colors are accurate and pleasing. I particularly like his pencil drawings, which, in the very palest of grays, manage quite astonishingly to catch the luminosity of a rose blossom and the intricacy of its buds, leaves, and thorns.

Now that good flower painting is on the wane, it is not surprising that we are turning back to the Old Masters of the art, who flourished in the golden age of botanical art—1730 to 1830. What *is* surprising is to stumble on a master of this period one did not know of, which I was recently lucky enough to do. Presumably the name and paintings of Moritz Michael Daffinger are entirely familiar to most Europeans, for he was a noted miniaturist in Vienna in the first half of the nineteenth century. Doubtless, too, his less well-known aquarelles of wild flowers are prized by most Austrians. I never, however, have happened to see them mentioned in an English or American book. They are not reproduced or even listed in Sitwell and Blunt's huge bibliographical record, *Great Flower Books —1700–1900*, possibly because they have never been published in a well-produced book. I came to know of Daffinger because a friend who lives in Vienna brought me two of his flower prints made from water colors now in the possession of the Akademie der Bildenden Künste, in Vienna. Some of the five hundred flower paintings are privately owned, some are lost; most of them were bought from his descendants by museums. My two prints are not large—they measure fourteen and a half by eleven and a half inches, including the white background—but neither were the wild flowers he painted. One print is of two wild anemones, *Anemone grandis* (lavender to violet) and *Anemone vernalis* (white). The entire plants are shown—flower, leaves, roots and all, exquisite in the detail of delicate root structure and slender green leaves, brown stalk, and gold center of the half-opened flowers. It is as if you had just dug up the two intertwined plants yourself in the woods and brought them into the house. The other is of a *Schneerose* (we call it the Christmas Rose), the white-flowered, gold-centered *Helleborus niger*, showing blossom, unopened bud, and stalwart green peonylike leaves. In spite of the careful detail, these are not literal illustrations for a botany; the instant you walk into the small room where they hang, you know you are in the presence of works of art—portraits of flowers. Though the colors are quiet ones, they glow from the wall.

Thanks again to my friend, I am now in possession of a second-hand copy of a book in German full of reproductions of Daffinger's art from beginning to end—painting on china, miniatures, costume designs, and flower paintings—with a brief biography of Daffinger by Emil Pirchan. The book was published by Verlag Wallishausser, Vienna and Leipzig, in 1943. Daffinger's life was a strange one. Born in a poor section of Vienna in 1790, he was the son of an impoverished porcelain painter in the Imperial and Royal Porcelain Factory. The father died when the boy was six, and although his mother soon remarried, she had to eke out the family income by doing fine embroidery for the noble ladies of Vienna. From his earliest childhood, Moritz Daffinger drew pictures, and by the time he was ten he was designing delicate patterns of plants and flowers for his mother's embroidery. This, and his work as apprentice in his father's factory, to which he went at the age of eleven, led to his being given a scholarship to study at the Imperial and Royal Academy of Art. He studied for seven years, and he stayed on in the factory for eleven years, where he continued until he was commissioned to paint a miniature of the Empress on ivory. This launched his career as a miniaturist, and he plunged into a whole new world—of the Court and of the theatre, in which he was much interested and for which he designed both costumes and sets. When he was still in his twenties, he fell wildly in love with the famous tragedienne Sophie Schröder. They never married, but she bore him two sons. A decade later, his friend Franz Grillparzer, the dramatist, introduced him to the beautiful Marie von Smolenitz, whom he married when he was thirty-seven, and by whom he had a daughter, Mathilde. When Napoleon's court painter, Jean-Baptiste Isabey, came to the Congress of Vienna to do miniatures of the assembled notables, Daffinger sat for his portrait in order to study Isabey's special techniques for increasing the luminosity and depth of flesh tone painted on ivory—techniques he was to adopt himself. Soon he became the most sought-after miniaturist in Vienna, and he demanded and got high prices. He would refuse to paint a face that he thought ugly or that did not interest him, and he was noted for his bad manners and high-handed ways; he even dared to reprimand members of the imperial family if he considered their conduct or dress improper. Perhaps he was already tired of the finicking art of the miniaturist, but it was the death of his daughter, at the age of fifteen, that gave his life a whole new turn. His sorrow was so great that he fled to the woods, where he roamed alone for days on end. Finally, he began to paint what he saw there—small flowers, butterflies, beetles—and for the first time he lost all interest in material values and large earnings and became friendly and mild-mannered, at last at ease with the world. In the next five years, he painted his five hundred water colors of wild flowers. In 1849, he died, a victim of the cholera epidemic in Vienna. The illustrations of the Pirchan volume concentrate on the miniatures, of which he left a thousand. Many of them are rather florid and fancy, and some seem to show a distaste for his subject, as Sargent occasionally did; others, particularly those of children and of his wife and daughter, have great charm.

The Pirchan book includes five not very good reproductions in color of his flower paintings, and sixteen in monochrome, but even in poor reproduction the pictures of such wild flowers as hardy cyclamen, primrose, gentiana, and saxifrage make me long for an album of the best of them, well reproduced in full color. It occurs to me that Daffinger's unhappy years of painting miniatures with a fine-pointed marten-hair brush on ivory leaf smoothed to a high polish with pumice, and of catching the luminosity of flesh tones or the glint of an eye or a jewel, may be in part responsible for his success in painting in water color the tiniest detail of a small flower and making the whole plant live and glow on paper as it had in the woods.

The beloved wild-flower book of my childhood, Mrs. William Starr Dana's *How to Know the Wild Flowers*, has become a classic, and at last it has been revised and brought up to date on nomenclature to conform to the latest edition of Gray's *Manual of Botany*. It has now been published as a paperback by Dover Publications. The Dover paperbacks are long-lasting because of their stiff covers, their good paper, and their sewing, which is like that of any well-made hardcover book. Bindings will not split or pages fall out, as do those of the average glued paperback book. Marion Satterlee's excellent line drawings of the original edition are retained, and so is all of Mrs. Dana's text except for a few passages that had become obsolete. One of the pleasures of the book is the lively literary quality of the writing and the author's supplementary material on such matters as the uses of the plants in the past and the myths attached to them. Even in late years I have often been able to identify a wild flower in the original Dana book, published in 1893, far faster than I could in the more modern ponderous and scholarly tomes. This new edition is the flower book par excellence to give a family of children. Mrs. Dana, if popular and old-fashioned, was no slouch as a botanist. Her small botanical summaries, given in italics at the start of each entry, sum up every feature a child needs for fairly accurate identification, and her introductory chapters explaining botanical terms and describing the "notable plant families" are instructive and clear. Clarence J. Hylander is the appreciative editor and reviser.

A beautiful but highly specialized book, yet to me a most appealing one, in its Oriental version of our garden "objects of delight," is *The Art of the Chrysanthemum*, by Tameji Nakajima, "with the collaboration of H. Carl Young." Obviously, though, Mr. Nakajima, with his long beard and ascetic, spiritual face, who in the photographs has the dedicated look of a Buddhist priest, is the genius of this unusual book on the art of growing and training chrysanthemums. The volume, published by Harper & Row, was printed in Japan, which may explain how it can be sold for less than ten dollars in spite of its handsome format, its forty-four plates in delicious color, and its many more black-and-white photographic plates, drawings, diagrams, and decorative "spots." These last are based on traditional Japanese family chrysanthemum crests, some of them belonging to members of the imperial family. (The Emperor's chrysanthemum crest is omitted—I suppose out of deference.) The first section of the book is a treatise on how to raise superior chrysanthemums both indoors and out. The second, and most interesting, concerns the intricate art of bonsai as it applies to chrysanthemums. The third explains how to train the larger potted styles that make Japanese chrysanthemum shows a marvel to Western eyes. I think these showy styles—"cascade," "cursive," "apron," and "thousand-bloom"—too artificial and even sometimes quite hideous, but if I ever had the time and patience and the chance to stay in one place long enough to attempt a bonsai (one can't desert a bonsai for even a day), chrysanthemums would, on the basis of this book, be my preferred medium. I would also give much to own one of Mr. Nakajima's tiny, crooked chrysanthemum trees covered with miniature blossoms, or one of his wind-swept chrysanthemum forests in full flower in a shallow dish. He is a revered master and teacher of the art of the chrysanthemum, and his collaborator, Mr. Young, a native of Iowa who stayed on in Japan after serving in the Korean War and studied for fourteen years the Japanese methods of chrysanthemum pot culture, has himself become a well-known teacher and exhibitor in Japan. It is to him, I suppose, that we owe the aids the book offers for American readers, such as a glossary of terms, an appendix listing nurseries in this country that supply chrysanthemums, and an introduction that, among other things, compares the climatic conditions of Japan and the United States. He doesn't, alas, tell me how to winter-over chrysanthemum plants in the cold temperatures of Maine (my luck has been about ten per cent), and he does not even try to tell us what American chrysanthemum cultivars come nearest to the Japanese-named varieties shown in the illustrations. ("Cultivar," by the way, is a coined word meaning "cultivated variety." I have always disliked the word, but it is now in such general use by horticulturists and writers on floriculture that I have had to succumb to it.) Any amateur gardener who grows chrysanthemums will find Part I useful, what with its cultural directions and disquisitions on soils, pests, and techniques like "pinching back." I was astounded, for instance, to discover that half the time the Nakajima bonsais grow with their roots largely exposed, and I remembered wryly how many times I have patiently covered my chrysanthemum roots with soil, doubtless to their detriment.

The City Gardener, by Philip Truex (Knopf), is by all odds the most complete and useful book I know for those who are ambitious enough to attempt roof, balcony, or back-yard gardens in any metropolis, and even for those who have to settle for a mere window box or for a pot-and-tub garden on a penthouse porch. The author is well equipped to give practical advice, for he has been through it all himself on his own roof. Also, he is the owner of a shop called the City Gardener, which for a decade has catered to New York's gardeners, those most beset and persistent of all amateur cultivators. He writes with enthusiasm and he writes well, and there isn't a stone or a plant or a contingency he has left unturned. I have attempted with only meagre success to garden in New York—in back yards and on porches and balconies—and though sootfall, gas fumes, and too much shade are now in the past for me, I wish I could have had Mr. Truex as my guide while I was trying it all. His book is in three sections: roof gardening, back-yard gardening (the two are totally different), and a twelve-month almanac. There are lists of city-tolerating plants, diagrams, photographs, a selection of recommended garden books, and an index. "Indispensable" is the best adjective for this readable volume.

For the suburban gardener, Donald Wyman's *The Saturday Morning Gardener* (Macmillan, 1962) gives equally sage advice. It is a guide to easy maintenance by the Horticulturist of the Arnold Arboretum, who also draws on his experience as a home gardener in a Boston suburb, with the help of his wife and teen-age sons. He covers planning, short cuts of many sorts, labor-saving machines, and so on, and he describes nearly a thousand low-maintenance plants, vines, shrubs, trees, and ground covers. Mr. Wyman warns us that plants and varieties that have proved difficult or easy for him may, in a different soil or locale, be the reverse, and this checks out with my own experience. Missing from his lists are three of the mainstays of my borders in Maine—hybrid columbines, delphinium, and Sweet William. The long-spurred columbines will renew themselves here at the drop of a seed, and the parent plants seem to go on forever. Modern delphiniums, if treated like biannuals, and good old Sweet William have given me next to no trouble. On the other hand, I simply can't seem to make hardy asters, and several other perennials that Mr. Wyman finds easy, do a thing. I might add that in my experience not very many Saturday gardeners have such cooperative young sons as Mr. Wyman seems to have.

A Gardener's Book of Plant Names, by A. W. Smith (Harper & Row), is a sort of dictionary, but it is much more than that, and it is by far the most entertaining garden reading I've met up with in years. In a way, it came as an answer to prayer, because I have long been interested in the origins of plant names and have searched in vain for any fairly complete book on the subject. Here are at least a thousand listings of plant names and botanical Latin descriptive adjectives—not a complete list, by any means, but a serviceable one. The author died just before the publication of his book. If he had lived, he might have given us another volume. Colonel A. Williams Smith was, like most educated Englishmen, a Greek and Latin scholar. His long military career took him all over the world and his travels gave him an interest in plants. He served in Russia with the Cossacks during the First World War and in Australia and Africa during the second one. And for a while he managed a teak forest in Upper Burma. Happily for us, he settled in New England on his retirement from the Army, became an American citizen, and raised fruit trees, vegetables, and flowers and made a collection of native wild flowers, shrubs, and trees. His book is American in slant, with English garden knowledge as a background. The entries list generic and specific names in alphabetical order, identify the roots of the botanical names—whether Greek, Latin, American Indian, or whatever—and, if they are not self-evident, explain the origins of the English common names. Most of the entries are brief. The charm of the book is that Colonel Smith, when he feels like it, takes off in full flight and writes an anecdote or even a sizable essay if the history of a tree or plant or the derivation of its name catches his fancy. The entry ROSE occupies two pages, and so does APPLE. I find it amusing to learn that Lycoris, a summer garden bulb, was "named in honor of a lovely Roman actress of that name, the mistress of Marc Antony"; that Tithonia was named for Tithonus, a young man much loved by Aurora, the goddess of dawn; that Tulipa (tulip) was the Latin version of the Arabic word for turban; that Calendula, which the English call "pot marigold," derives from *calendae*, the Latin word for the first days of the month, on which interest must be paid, and that its English name comes from the fact that in early English cooking a single head of the flower was thrown into the pot to flavor thick soups. Myths, historical facts, and etymology are thrown into Colonel Smith's pot in a savory mixture. Under PINE, he reminds us that in the seventeenth century all the noble New England white pines whose trunks measured twenty-four inches or more were marked as Crown property and were cut and transported to England to be used as ships' masts, and that as a consequence it is seldom that a pine board wider than twenty-three inches is found in early New England houses. He reminds us, too, that this confiscation was one of the many causes of dissatisfaction leading to the Revolution. Franklinia (named for old Ben), that rare and beautiful shrub with an odd history, does not get a separate listing, except as a variety of Gordonia, but George Washington is represented by Washingtonia, "a tall, not very handsome fan-palm native in California and the southern United States." Poor George! We should have done better by him botanically. Colonel Smith's essay on APPLE is particularly agreeable. I'll bet you did not know that the first named variety in America, Blaxton's Yellow Sweeting, was introduced around 1640 by the Reverend William Blaxton, who had a fruit orchard at what is now the corner of Beacon and Charles Streets, in Boston. "One of the best gardeners of his day, he also trained a bull to the saddle and used it to ride around Boston on his daily business." I could go on and on, but you will have more fun reading the book yourself. Some of the reviews in the botanical and horticultural journals have pointed out a few minor errors in Colonel Smith's nomenclature, and one of them even complained that the entries were of uneven length, which seemed to me to miss the whole point of this unusual book.

I am not much of an herb grower, if only because my family prefers its food flavors undoctored. Therefore, the culinary herbs I raise have been of the simplest sort—mint, parsley, tarragon, and chives. These all used to occupy a small kitchen garden devoted mainly to head lettuce, but some of them are now wandering or missing. The two or three varieties of mint growing there threatened to take over the whole bed, so they were moved just outside the home-ground fence into a hayfield, with the result that if I forget to warn the haymakers not to mow them down, as happened this year, our mint supply is temporarily sparse. Our chives proved so attractive to one of our dogs they were temporarily banned from salad. (We have no dog at the moment, but I have just discovered that some painstaking yet unacknowledged villain appears to have spirited away my two big clumps of chives.) Years back, I had some borage plants in this same bed, given to me by a friend, but I was never able to find a use for this tasteless herb, and it, too, was moved from the kitchen bed into the hayfield, where it still flourishes among the grasses. I would have thrown out the stuff long since if my friend had not kept saying, "Borage for courage, you know, borage for courage," and I've always felt I could use a little extra courage. Before the frost this fall, I went out to the hayfield and ate a sprig of borage to check my memory—no taste. Possibly my friend's plants weren't borage at all, but they do look like its pictures.

Because of this sorry record as an herb gardener, I hesitate to recommend two recent books on herbs that I admire and that may have made me more herb-minded. I have renewed the mint in the kitchen bed and put in two small plants of thyme—one English thyme and one French thyme, obtained through Merry Gardens' admirable mail-order catalogue from Camden, Maine. And in the house, from the same nursery, I am starting for next summer's garden a pot of lemon verbena and a pot of rose geranium for fragrance. I have grown these before, but not of recent years. Nothing is more delicious than a lemon-verbena sachet for one's bureau drawer, or a rose-geranium leaf to pinch as one walks past it in the garden.

Margaret Brownlow's *Herbs and the Fragrant Garden* is, as its title implies, a book on fragrant plants and shrubs as well as on culinary and medicinal herbs. It was first printed, privately, in England in 1957, then enlarged and revised and republished there in 1963. The American edition (McGraw-Hill) is a large yet portable volume illustrated by photographs, charts, and plans, but chiefly by the author's thirty-two pages of delicate water colors, each page containing many varieties of herbs or flowers. Eighteen varieties of thyme, for instance, are shown on a single page, but the effect is not crowded and the page has a pleasing pattern. Miss Brownlow paints with a sort of innocence and clarity that make for easy identification, yet the illustrations are those of an artist rather than of a botanical illustrator. To me, they far surpass photographs in their appeal to the eye and in their usefulness for identification. The text is equally pleasurable and useful. Its chapters include the fascinating history of herbs, advice on laying out and growing an herb garden and on harvesting its products and using them in cooking, salads, potpourris, and much more. A reference section, which occupies more than half the book, gives a table of flowering dates, short classified lists of aromatic and scented shrubs and North American herbs, and a long alphabetical list of the same with descriptions and advice on each plant, warnings against poisonous varieties, and much lore and history of the various species as well. There is an index. In spite of the book's orderliness, the writing is literary and personal, with poems, quotations, mythology, and personal experience interspersing the practical advice and the botany and horticulture. Perhaps I like it so much because Miss Brownlow appears to have my own predilection for fragrance and aroma, though I believe my pleasure comes chiefly from the author's vast knowledge and her patent enthusiasm. She never gushes; she merely enlightens the gardener and enlivens the reader.

The other herb book I like is an oddity titled *Garden Spice & Wild Pot-Herbs.* It is a collaboration between Elfriede Abbe, the illustrator, and the writers of the text, Walter C. Muenscher, late Professor of Botany at Cornell, and Myron A. Rice, a horticulturist and formerly a research scientist at Yale. The first edition, which appeared ten years ago, was limited to ninety copies, and for these copies Miss Abbe set the type and printed it on a hand press; she also designed the book and made the bold woodcut illustrations. The Cornell University Press later the same year brought out a popular edition, which was an exact reproduction by the offset process of the handmade book. Now the Cornell Press has reissued the book, boxed, perhaps because it is such a contribution to graphic art, in which field Miss Abbe has acquired considerable renown. I do not happen to like the type in which she chose to set her handsome volume—Goudy Kennerly Bold & Italic, a really gigantic type face, which is decorative yet which I feel has the appearance of a rather fake antiquity (most of the "and"s are ampersands, which leap out at you alarmingly from the page)—but I do greatly admire her vigorous woodcuts, which are stunning examples of graphic art and are also, I assume, completely accurate botanically, since the versatile Miss Abbe holds the post of Scientific Illustrator in the Department of Botany at Cornell. Designing and illustrating fine editions appears to be her hobby, and, to cap it all, she is a sculptor and woodcarver of note. (A busy lady. This year she turns up again as the author, illustrator, plant researcher, and translator of Vergil in another Cornell University Press book, *The Plants of Vergil's "Georgics."* I found it dry; she appears to have little interest in the *poetry* of Vergil. However, it represents vast research, and classicists, medievalists, and botanical historians will doubtless welcome it.) To return to *Garden Spice & Wild Pot-Herbs*, the text is thoroughly enjoyable, what with its exact botanical descriptions of two hundred and fifty plants and its suggestions on how to find and grow them, and to me it is particularly so because the two professors appear to have quite as lively an interest in cooking as in botany and horticulture. Recipes are given, and they sound good. I especially like the section on potherbs, most of which we commonly think of as weeds. Have you ever tried the first shoots of milkweed, cooked like asparagus, or made a Scotch nettle pudding, or eaten purslane (pussley) the way it is cooked in Europe, with butter, zwieback crumbs, meat broth, and cream or the yolk of an egg? Even pigweed can be a delicacy. Buy the book and see how.

Joseph Wood Krutch's *Herbal* is another decorative book of exceptionally good design and typography. It is not a book about herbs for gardeners or cooks, as the two preceding books are, but is Mr. Krutch's attempt to write his own medicine and plant book, or "herbal," in the manner and spirit of the classical and medieval herbalists, ranging from Dioscorides the Greek to the pre-Linnaean herbalists, whose books were a weird compound of early botany, mythology, astrology, and beginning pharmacology. It is likely to fascinate anyone who has not delved into Mr. Krutch's sources in a library, or who has a taste for quaintness and an interest in the dark ages of medicine, or who likes to marvel at the comic superstitions and occasional good sense of our ancestors. Although it is a work of scholarship and written with style, I think the "wit" ascribed to the author on the book jacket is in this case mostly a matter of the reader's laughing *at* the odd beliefs of Dioscorides, Pliny, Theophrastus, John Parkinson, Nicholas Culpeper, and the others. The big volume is illustrated by the curious woodcuts of

Pierandrea Mattioli's *Commentaries on the Six Books of Dioscorides*, issued in Prague in 1563. They are handsome examples of the early plant illustrator's art, and the choice of plates seems to have guided the author's rather arbitrary selection of the hundred plants he covers. Some of Mr. Krutch's passages on modern drugs, many of which were used in medieval times, are interesting, but aside from this, the book seems to me to come dangerously near to belonging in the conversation-piece category.

Quaint superstition and early medicine enter into another book—a small one this time, called *A Witch's Guide to Gardening*, by Dorothy Jacob, an English author, which in a wandering way covers much garden lore and legend and explains the Devil's hold on the natural world, especially in Great Britain. If witchcraft is your meat, this may be your book. I enjoyed the chapter on tree magic, much of it new to me, but on the whole the best thing the book did for me was to send me back to John Donne, whose magic and mystery I find more compelling:

Go, and catch a falling star,
Get with child a mandrake root,
Tell me, where all past years are,
Or who cleft the Devil's foot . . .

If you remember, Donne, in the same stanza of his song, wants to be taught what wind "serves to advance an honest mind." It is a wind every writer must search to find—especially, perhaps, every writer or artist dealing with that "idea which we have come to call nature."

A number of years ago, I expressed in these pages the hope that an American publisher would bring back into print some of Gertrude Jekyll's garden writing of the late nineties and early nineteen-hundreds in an anthology selected from her thirteen books. Last year, Scribners did just this, in a volume of selections titled *On Gardening*. It is a handsome book, dignified and tasteful in typography and format and discreetly illustrated with a few strong pen-and-ink drawings by Margaret Philbrick. (The compiler is not named.) One feels that Miss Jekyll, an artist herself, would have approved of her new book but might have felt that a condensation of her two best books would have been a wiser plan. The selections are well chosen and are as useful and interesting to gardeners today as they ever were, but to the Jekyll addict, anyway, they seem a bit skimpy and scattered. I, for one, missed the continuity of the tremendous tale Miss Jekyll unfolds in her first two books—*Wood and Garden* and *Home and Garden*, in which she describes the making of her famous garden and the building of her house at Munstead Wood. The preface of the new book, by Elizabeth Lawrence, briefly and brilliantly sums up Gertrude Jekyll's life, describes her quirky and interesting personality, and assesses her contribution to the landscaping, gardening, and horticulture of her period and of ours. Her portrait is a useful antidote to recurrent snipings at this forceful lady's spinsterish habits and so-called plain looks. She was gallant and she had the good looks of character and vigor. In her fifties, giving up her work as an artist because of failing eyesight, she was able to create a wholly new career, in which she became an authority and an influence both in England and in this country. "In an age of fine writing and fine gardening," says Miss Lawrence, "Gertrude Jekyll excelled in both." Except for Miss Lawrence, I can think of almost no one of whom this could be said today.

Another exception could be Mrs. Mortimer J. Fox (Helen M. Fox), a knowing garden writer and gardener, to whom I am indebted for an interesting footnote on Miss Jekyll's old age. (She lived to be eighty-nine.) In 1960, Mrs. Fox wrote me, "Years ago I visited Gertrude Jekyll. She was too ill to see me but told me to visit the garden. While I was there, along came two maids with their streamers floating from their caps and their blue-and-white uniforms covered with large white aprons. They were carrying a large wash-basket which they held under one rosebush after another, and tapped the roses so the petals would fall into it. It was a delightful picture and so very old world." It should not be assumed, I feel sure, that this episode was the result of an old-maidish tidiness, since Miss Jekyll was always for naturalness and disliked a fussily neated-up garden. What was going on that day must surely have been the gathering of rose petals for Miss Jekyll's famous potpourris, described in both *Home and Garden* and the new anthology. It is pleasant to think that, old and ill though she was, Miss Jekyll, if only vicariously, was attending to her annual rose harvest, which would bring her garden indoors and preserve its ineffable June fragrance through the long winter months to come.

9

The Million-Dollar Book

Anemone grandis *and* Anemone vernalis.

By Moritz Michael Daffinger, 1790–1849

December 10, 1966

Every generation or so, a monumental work in the field of botany is undertaken, and this year we have the start of such an enterprise in *Wild Flowers of the United States*, sponsored by the New York Botanical Garden, under the supervision of Harold W. Rickett, and published by McGraw-Hill—two huge and beautiful volumes for popular consumption, which are but the first of a series of ten. The others will follow, two at a time, at intervals of a year or less. Before I get to the new books, I would like to look back three or more long generations to Charles Sprague Sargent's *Silva of North America*, another massive publishing event, on which work started in 1882 and was completed twenty years later. The first of its fourteen folio-size volumes was published in 1891 and the last in 1902. In this strictly scientific work, Professor Sargent, America's great dendrologist, set out to list and describe in botanical detail every species of tree in North America (exclusive of Mexico), with complete synonyms, references to literature, and copious notes on the economic, horticultural, and general value of each tree. The Smithsonian Institution had estimated that the work would take seventy-five years and offered to back it, but this long-term plan did not satisfy Professor Sargent, who engaged the gifted botanical artist Charles Edward Faxon to make the illustrative line drawings and see through the making of the seven hundred and forty plates in France under the supervision of two French botanists. The cost of the *Silva* in dollars and cents cannot now be determined in the set fashion of this year's *Wild Flowers*, which was first estimated to cost a million dollars for the entire series, but the years of exploration, scholarship, and effort that went into *Silva* can be guessed at in a dim and awestruck way.

As soon as it was completed, Professor Sargent began a careful condensation of *Silva*, and this shorter version, titled *Manual of the Trees of North America*, is far better known and more useful to most of us than the great scholarly work. Even the *Manual* was labored on over a period of twenty-two years. The first edition was brought out in 1905, with the Faxon plates. A second edition, enlarged and corrected, with forty-three additional illustrations by Faxon and Mary W. Gill, was published in 1922, and this edition, with corrections, was brought out again in 1926. This final edition is fortunately still available to us in a paperback version in two volumes put out only last year by Dover Publications. The Sargent text and the original plates are untouched, but an appendix has been added to bring the nomenclature up to date—a necessary curse and a blessing, since the busy taxonomists never stop fussing with botanical Latin names and plant classification. Although the text is written primarily for botanists and horticulturists, these two handy and not overlarge volumes are still the best reference books for the amateur to consult if he is faced with the identification of a tree or the decision on what trees to plant in his climate zone. In his preface to the second edition of the *Manual*, the author writes that it contains the results of forty-four years of his continuous study of the trees of North America—a phenomenon of faithful dedication. Yet after all, the book was just one of his side efforts. Harvard's Arnold Arboretum was Charles Sargent's outstanding accomplishment.

I think of Professor Sargent with veneration for his vast achievements, but I think of him also with affection, not because I ever knew him beyond shaking his hand once or twice but because he was a stimulating and generous neighbor during my childhood. Holm Lea, his great estate of a hundred and eighty acres, was within walking distance of both the houses my family occupied in Brookline, and it was traversed by driveways and paths, which he indulgently allowed his neighbors' children to bicycle through whenever we wanted to. It was, of course, made plain by our parents that we were never to pick a flower or a branch on Mr. Sargent's land. Though we did not know it then, what we saw, as we biked, was a model of extraordinary home and estate planting which was to influence landscape gardening all over New England—indeed, all over the nation. Each year, too, on several Sundays in May and June, when his lilacs, his wisterias, or his rare collection of hardy rhododendrons and Indian azaleas were at their height of bloom, Professor Sargent opened Holm Lea to the town. My father seldom failed to go there once a year with the rest of the throng, and I usually tagged along, even when I was quite young, to walk the paths around the pond and see the soft, blended colors of the flowering shrubs reflected in the water, and occasionally to go up on the porch of the Sargent house, where some tenderer blooms, in pots, were on display. Professor and Mrs. Sargent sometimes were there, she smiling graciously in the background, he a tall, somewhat portly, and imposing bearded man with a commanding Roman nose, hovering protectively over his plants yet wanting the public to enjoy them at whatever cost of privacy to him and his family. He and my father would exchange neighborly if formal greetings, and very likely, in the earliest years of our visits, they discussed the eternal problem of the mixup in their mail, because for a short time both families lived on Walnut Street and the two names, "Charles S. Sargent" and "Charles S. Sergeant," were often too much for the postman. Professor Sargent also meant to me the Arnold Arboretum, in Jamaica Plain, to which we went by horse and carriage, to view the hawthorns or lilacs in bloom and study the names on the trunks of the rare trees.

The story of the Arboretum, which was put under Professor Sargent's care with only a few thousand dollars to work with when it was a run-down farm, and of his building it up to be one of the greatest institutions of its kind in the world, would make a book in itself. There was a battle to raise funds and a battle with the City of Boston to persuade it to incorporate the Arboretum into its park system. Professor Sargent won the incorporation by persuasion, and the city assumed the cost of building and maintaining the Arboretum roads and paths and of policing the grounds. The grounds were given to Harvard, which turned them over to its own Professor Sargent for planting a collection of shrubs and trees. Then followed the years of sending out plant collectors all over the world. E. H. (Chinese) Wilson, the most famous of these men, brought back to the Arboretum a rich collection of the flora of Eastern Asia. The Professor's own travels were far and wide.

Earlier still, Professor Sargent, when he was thirty-eight, headed a commission to study the forests of the United States in order to estimate our forest resources for the Tenth Census, of 1880—a report that was published in two volumes as an appendix to the Census. (How pleasant to have lived in a day when a count of trees was more important to the national economy than a count of motorcars or television sets!) The report resulted in the establishment of a rudimentary national policy on forests—perhaps the very beginning of our concern to save the wilderness areas. Thus it seems to me that an even better book would be a biography of this great and energetic and sometimes forbidding man, to tell the full story of his civic and botanical achievements, of which I have given only the barest hint. Little has been written about him. I have been able to turn up only a brief memorial article by one of his colleagues, Charles Rehder, in the *Journal of the Arnold Arboretum* (Volume VIII, 1927), and a chapter titled "Garden Adventure" in Gladys Brooks's poignant and well-written book of memoirs, *Boston and Return*, which is the only personal account of him I have come across. Mrs. Brooks, at the age of nineteen, when she was Gladys Rice, went to the Arboretum with a friend of the same age to study in preparation for what they hoped would become a career in landscape architecture, and they were lucky enough, in the later years of the Professor's life, to be taken on as unorthodox students and apprentices. Gladys Rice thus came to know the austere Professor in a very special way. After putting them through a gruelling test and a course of education with one of the Arboretum's gardeners, he accepted the two young ladies as friends and protégées, always referring to them as "the little girls." The sketch is charming, witty, and revealing—the only real portrait, for all its brevity, of this great man. The author sums up his personality far better than I can:

> The obduracy and stamina, however, tied to a high disposition of mind, were attributes that produced the enduring monuments Professor Sargent was to leave behind, a well-woven rope from which are suspended the grandeur to be seen in forest planting, in the parkways of our land, along the avenues of great cities, and in the enhanced beauty of private gardens: a far, fair world.

The magnitude of the work of compiling *Wild Flowers of the United States* is comparable to that of *Silva* and its offspring, the *Manual*, but the reasons for the size of the two tasks are different. When Professor Sargent started his study, not all the native trees of our continent had been discovered, named, listed, and classified. His two books, therefore, were to some degree pioneer ventures in plant discovery. His final edition of the *Manual* includes sixty-five genera and seven hundred and seventeen species of trees. When Harold W. Rickett, a senior botanist of our Botanical Garden, who wrote the text of *Wild Flowers*, began *his* labors, with six collaborators, the wild flowers of the continent were already known and fully described in scientific publications. But the numbers of our flowers of forest, meadow, swamp, prairie, littoral, and mountain far exceed the number of our trees. The wild flowers of the Northeastern states alone, which are those covered in these big first volumes, number three thousand, of which seventeen hundred are described and twelve hundred shown in color plates, and three hundred and fifty in line drawings. Before the series is completed, in ten big books, Dr. Rickett estimates that between ten and fifteen thousand species will have to be considered for listing—a staggering prospect.

The story of the great undertaking has been told already in this magazine. William C. Steere, director of the New York Botanical Garden and supervising editor of the great tomes, tells it well, in all its drama, in his foreword to Volume I: the raising of the million dollars, the benefactors, and the problems of selection and production. I shall repeat only, since it seems to me important, that the impetus for the venture came from Diarmuid C. Russell, a gifted literary agent, whose enthusiasm for gardening and for wild flowers equals his dedication to the cause of good writing. He, after all, had the *idea* for the book and the energy to interest others in his idea. Ten years ago, as Mr. Steere tells it, Mr. Russell "had the perspicacity to realize that not only were most popular books on American wild flowers incomplete and undependable but also that they left great geographical areas untouched." We amateurs of botany will always be in his debt, since the book most nearly comparable lists only five hundred flowers. We also owe a debt to Dr. Rickett for his simple, understandable, down-to-earth text, which avoids technical terms wherever possible but does not distort by popularization or duck botanical terms where they are needed. The Latin names and the common English names are given. Not quite all the regional names, though. He omits, for instance, any listing of the much used but misused common name "cowslip." To New Englanders, marsh marigolds are cowslips; to Virginians, bluebells are cowslips. To the English, one of the primroses is the cowslip, and this is perhaps the only correct use of the common name, yet often in my childhood bicycle rides through Holm Lea my friends and I entered at the Walnut Street entrance because it was a short cut to Heath Street and a lovely swamp where what we called cowslips and skunk cabbage grew. We picked them, even the smelly cabbage, and took them home in triumph. My cowslips were marsh marigolds, but I had never heard that name then. They were cowslips to me, and perhaps they always will be. In the listings, the common name sensibly comes first, but what is a reader to do who knows a flower only as a cowslip?

The Rickett preface is a model of clear exposition on the numbers, nomenclature, and identification of wild flowers and on the botanical composition of flower, fruit, stem, and leaf. Even more valuable to me, who have always been befuddled in books listing flowers by families, is the "Guide to the Families of Flowering Plants," with a clarifying chart for help in their identification. There is also an illustrated glossary of botanical terms.

Here I should explain that the two books published this year are called Volume I, Parts One and Two, which I find bothersome. After all, they *are* two volumes, both bearing the subtitle on the spine of "The Northeastern States," and I see no reason for not calling them Volumes I and II. I hope to live long enough to own all ten volumes of the completed series, and I know I'd find them easier to shelve and to consult if they were simply numbered I to X. The index for the two parts is at the end of Part Two, which is doubtless the reason for all the fussiness, but that same index for the Northeastern states should have been included in both volumes.

The books owe much of their good looks to the color-photograph pages. Perhaps the biggest job of all was to obtain and select the photographs and get the color plates made in England. The color is nearly always true to nature, and the photographs are straightforward ones, without any distracting or fancy effects of special lighting or background. Fifty-three photographers are represented, among them Dr. Rickett and, notably, Charles Johnson and Samuel Gottscho. Each volume has only one full-page photograph, and these two are the loveliest of all, because the size helps place the flower in its setting—the delicate scarlet cardinal flower in Volume II, growing in damp woods, just as I used to see it in my girlhood in New Hampshire along the borders of the brook that was the outlet of Chocorua Lake, and the showy rose mallow in Volume I, on the edge of its usual watery marsh. For the most part, the photographs are arranged seven or eight to a plate (or full page), with lots of white surrounding them; the Latin name of the flower and the name of the photographer appear in small print underneath. I found annoying the absence of a cross reference on the plates; each should tell the number of the page on which the text for the flower is to be found. (The texts, on the other hand, do contain references to the plates.) As a consequence, if an amateur's first identification of a flower is by the picture, which is only too often the case, he must turn to the index to find out what the common name is and on what page the flower is described and analyzed. The description, according to the Botanical Garden's prospectus of the book, is "always located within a page or two of the photograph," but this is by no means always true. Often picture and text are eight pages apart, and in the case of the marsh marigold the text is on page 120 of Volume I and the photograph is on page 521 of Volume II. Sweet flag, maddeningly, is separated from its photograph by four hundred and thirty-five pages. Quite a trip to make, particularly since the two books are so big that it is inconvenient to have both of them on an average-size desk at the same time. (As I write, one of my volumes lies open on my couch and the other lies open on my desk.) Whipping back and forth between picture, index, and text is dizzy-making, which is why I wish that the index had been printed at the end of each of the two big volumes and that every illustration carried the page number of the descriptive text. I wish even more—although it would have made less handsome books—that the color illustrations had been inserted right opposite or in the midst of the pertinent text, as botanical drawings usually are. In other words, I must report sadly that I do not find these handsome books easy to use as reference books. Their size, ten inches by thirteen, and their great weight are a drawback. Although the size gives a chance for beautiful photography and large, easy-to-read Caslon Old Style typeface on heavy paper, I would have settled for a smaller page and smaller type and far less elegantly heavy paper. My impression is that the publishers wanted to produce two or three things at once—a comprehensive reference book for amateurs, a luxury picture book for the library shelf, and, as a McGraw-Hill editor said to a reporter, a book that would be "the Audubon of wild flowers."

The Audubon of wild flowers? *Wild Flowers of the United States* does not achieve this by a country mile. Audubon's *Birds of America* was the work of one man's lifetime—a pioneer work of ornithology to discover, identify, and study the habits of the birds of what was then a partly unexplored continent, and to illustrate them in color. Most important of all, the illustrations were prints made from Audubon's own brilliant paintings, combining water color, pencil, ink, chalk, even oil. This year's new edition of Audubon, published by Houghton Mifflin and American Heritage, reproducing not the familiar prints but the hitherto unknown original drawings and paintings, places Audubon as one of our great early artists. Their horticultural backgrounds, too, some by Audubon but more by his assistants, are often masterpieces of botanical art, particularly those by his son, John Woodhouse Audubon, Joseph Mason, his young apprentice with whom he afterward quarrelled, and the later ones by George Lehman and Maria Martin. As for me, I wish that the initial million-dollar investment in *Wild Flowers* could have been doubled in order to illustrate the big volumes by the paintings of what few master flower painters remain to us.

Color photography, for all its increasing merits and beauty, does mislead quite often; the camera is not always a truthful reporter—far less so, I honestly believe, than the artist. There is the matter not only of the reproduction of the colors, which here miss only occasionally (usually in the blues), but of getting topnotch color film to start with. The azure or bright-blue fringed gentian comes out here as a soft violet, and the brilliant bright-blue blossoms of bugleweed (*Ajuga reptans*) are sadly blackened. Incidentally, the photograph gives no hint of this variety's snakelike creeping habit, which explains the Latin adjective. There is, even more important, the matter of scale. See the photograph of the tiny bluets that in spring pattern our green pastures and meadows with their blue-pink-white sheets of blossoms, a half inch or less in diameter. As photographed here, they look almost as big as Dutchman's Breeches. This is because the scale of the photographs follows no rigid rule. All the drawings, by contrast, in the *Manual of the Trees of North America* (except a few that are clearly noted) are exactly one-half the size of nature, as are the three hundred and fifty useful line drawings in *Wild Flowers*. A consistent scale is a help to the amateur trying to identify a newly found flower. Photographs also mislead us in relation to detail. Take the yellow loosestrife, a common wild flower, whose several varieties are either clear yellow or yellow marked with dark dots or streaks. The dots or streaks do not show up in the photographs here, but they probably would have shown up in a plate made from a painting or even from a line drawing. On the other hand, I am indebted to the text on the loosestrifes. A few years ago, I noticed an unusual stand of golden spires by the roadside as I was motoring and picked a few to bring home to identify. As I remember it, they were clearly, from the leaves, the whorled loosestrife (*Lysimachia quadrifolia*)—not swamp candles (*Lysimachia terrestris*)—yet they looked unlike the photograph of either variety in *Wild Flowers*. Dr. Rickett, good botanist that he is, probably gives the answer to the mystery when he says the two varieties sometimes cross. Mine were doubtless hybrids, for, as I remember it, they had the inflorescence of the swamp candles and the leaves and spots of the whorled loosestrife. No other reference book I own told me this pertinent point.

In his editing, the author has had to make some hard decisions about what flowers to include and what to exclude. Trees, shrubs, grasses, sedges, rushes, and unattractive weeds with small greenish flowers unlikely to excite the interest of the amateur have been ruled out. This means no wild roses at all, no elderberries, no wild azaleas. However, there are some sensible exceptions. The wild spiraeas, such as hardhack and meadowsweet and some others, although rightly shrubs, are common roadside flowers, and their woody stems are not conspicuous. Therefore, they are in the book—beautifully pictured. Swamp laurel (*Kalmia polifolia*) under this system gets in because it grows very low, but not the best-known *Kalmias*, such as mountain laurel and lambkill. The leaves of laurel are poisonous if they are eaten, and this warning is given with the entry; poisonous plants are helpfully described as such throughout the book. I found it startling to learn that water hemlocks, the *Cicutas*, are among the most poisonous plants of this country, because the blossoms of spotted cowbane (*Cicuta maculata*) are almost lookalikes for Queen Anne's Lace, the wild carrot—at least in their photographs here. The same is true of *Conium maculata* (poison hemlock), which Dr. Rickett tells us was the plant used for the execution-by-poison of the criminals of ancient Greece, including Socrates. All these years I had been ignorantly visualizing Socrates drinking poison made from the needles of the majestic hemlock tree, which, for all I know, are not poisonous at all.

One is grateful for such bits of history and myth as are included, but they are rather few and far between, yet excellent when they are given. One is grateful, too, for the occasional derivations of and reasons for the Latin or common names. *Kalmia* is named for Per Kalm, a student of Linnaeus, who was sent by the Swedish government in 1748 to explore the natural resources of the United States. Dr. Rickett does not tell us this when he describes the *Kalmias*, but he does explain—many hundred pages and one volume earlier—that the botanical Latin adjective *kalmianum*, describing one species of *Hypericum*, is named for Per Kalm. He seemingly does not, however, have space to tell us that the common name for the *Hypericums*, St.-John's-wort, came about because the plants blossomed in England around June 24th, St. John's Day. That is to say, this is not a chatty book, nor should it be. The amateur who reads wild-flower books for armchair fun must turn back to Mrs. Dana's *How to Know the Wild Flowers* for legend, mythology, and poetry, and to A. W. Smith's incomplete *Gardener's Book of Plant Names* for the derivation of flower names. The unpleasant Latin adjective describing the beautiful five-foot-tall great blue lobelia (*Lobelia siphilitica*) is, nonetheless, interestingly explained by Dr. Rickett. The plant was said to be a secret remedy of the American Indians for syphilis. The secret was purchased from the Indians by Sir William Johnson in the eighteenth century, but, Dr. Rickett points out, the money was ill-spent. There is no mention, he says, of the plant as a cure in medical books of the nineteenth century, not even in the early ones.

I have learned many things from reading Dr. Rickett's text, some of them of import only to me, such as that the yellow hawkweed that is gradually taking over our lawn is the species known as mouse-ear—a far more agreeable name than hawkweed. I had never bothered to look up the pest before and simply called it hawkweed. The mouse's ears are the springy rosettes of leaves, and to me they have become acceptable, where necessary, as a substitute for lawn grass. We used to pluck, dig, and rake away at their spreading runners, but we failed to stem the tide, and now I am quite content even to play croquet on a well-clipped surface of mouse-ears. I have not yet been able to determine from the section on violets all the species names of our several regional violets. I suspect that our commonest, big, odorless purple violet of field and roadside is *Viola papilionacea*, the commonest violet of the Northeastern states, but we have another without fragrance that is very much like it. When spring comes, I can pick some leaves and compare them with the excellent line drawings of violet leaves that are added to help with identification. I shall also have to wait until warm weather to determine how important Dr. Rickett thinks the nose is when it comes to pinning down the species of a flower. He makes very little of fragrance except when it is outstanding, as in the wild white violet (*Viola blanda*) and the common red clover. Many wild flowers have no scent, or next to none, yet many not notable as sweet-smelling do have a subtle, unassertive fragrance. In all probability, the sense of smell is too uncertain and variable a matter for a botanist to bother with, yet the first thing I do is to smell a newfound wild flower. To me, it would seem a help if all reference books carried notes on fragrance. One mystery of wild-flower scent remains unexplained. Two of my granddaughters, when they were little girls and knew no better, picked by the side of the Maine turnpike on their way to visit me a few white Indian pipes—the ones Dr. Rickett names as *Monotropa uniflora*. The text says nothing about their fragrance or lack of it. Other books have told me that this species is without odor. The thing that hit me first was that the sad little pipes gave off a sickish sweetness. These were certainly *not* the pink Indian pipes (*Monotropis odorata*), which are the only Indian pipes Dr. Rickett notes as fragrant. The pink scented species does not grow in Maine, and, besides, these were dead white.

I should point out that the region covered in the two volumes includes more than most of us think of as the Northeastern states. It runs from Maine through Minnesota and Iowa, and from the Canadian border through Virginia, Kentucky, and Missouri. Therefore, a good many temperature zones are included, and a reader, like me, in Maine is faintly surprised to find that three yuccas, a genus characteristic of the Southwest, are included among flowers of the Northeast. A large map covering the endpapers of both volumes shows the whole country in green except for the region described, which is pure white—a happy device. The botanists have also been meticulous in pointing out in every case the wild flowers that are not native but came from the Old World or elsewhere, by one means or another, to establish themselves as naturalized citizens of our North American wild-flower population.

I believe strongly in regional handbooks of a small and manageable size. Therefore, my most earnest hope is that Dr. Rickett and his colleagues will not rest from their labors when they are finished with their ambitious series but will start immediately, as Professor Sargent did so energetically on the completion of his *Silva*, to boil down the size of their books (in this case, though, not the length of the text or the number of species) and reissue the work, divided up into handier and cheaper books, arranged state by state or by small clumpings of neighboring states. They might have to figure out how to illustrate these small books in a less expensive but still beautiful way. They could even consider the method—though I hope not the American edition's color reproduction—of *The Concise British Flora in Colour* (Holt, Rinehart & Winston, 1965), by the Reverend W. Keble Martin, in which thousands of wild flowers are described and nearly fourteen hundred are illustrated, in exquisite line and color, by showing many, many flowers on each color plate. Of course this would require a botanical artist of the calibre of the Reverend Mr. Martin, who is just the latest of the British clergy to emerge as a flower devotee and expert. Dean Hole and his *Book About Roses*, Canon Ellacombe and his *Vicarage Garden*, the Reverend Francis Kilvert, the young curate describing in his *Diary* the "Easter idyll" of the dressing of the graves with wild primroses in his Welsh village—we owe much to the Church of England over the years. For beauty and accuracy, the Martin book can't compare with *Wild Flowers of the United States*, but it is a handy, inexpensive, and attractive book for travellers in Great Britain. Few amateurs can afford the huge new beautiful million-dollar book.

Despite my strictures, I do indeed rejoice in *Wild Flowers of the United States*. The estimate for the series has already gone up a half a million, and very likely the cost will be more before the final volume is completed. Let us be grateful to those who made it possible. Meanwhile, I shall lie down in a wide double bed whenever I can with the first two gorgeous books, since I can't hold either one of them in my hand for long, and I shall prop them up in turn on a pillow and feast my eyes on the glowing plates of the goldenrods and *Hypericums*, and the exquisite ones of the violets and water lilies and all the other lovely and delicate denizens of our polluted and ravaged land. We must read about them fast and look long at their portraits and search them out in their haunts before they disappear, one by one, through the depredations of Man.

Why are wild flowers so important to those of us who care at all for flowers? For me, anyway, it is because they come like gifts from God (or Nature), and to encounter them in their natural habitat is an extraordinary aesthetic pleasure—such as trolleying all one Sunday morning with my father during my childhood to reach a hillside north of Boston where the fragrant mayflowers, more accurately called trailing arbutus, grew in profusion. We wrongly, as I now know, cut a handful to take home, and thus in our small way we may have helped toward the possible extinction of this exquisite flower. The flower peddlers who then sold them by the hundreds of bunches on the streets of Boston were the true local villains. (Dr. Rickett makes a plea for viewing, instead of picking, wild flowers, and he is right. We must learn to be like the Japanese, who go out to view, admire, and derive spiritual value from flowers rather than to pick them.) From my childhood, too, comes the excitement of finding my first wood lilies and lady's-slippers in a pine forest on Cape Cod, and my first fringed orchid in a New Hampshire swamp. Some favorite memories from my adult life are of walking between the high hedgerows that wall the narrow roads of Chagford, in Devonshire—hedgerows full of flowering briars, with countless delicate tiny flowers at their feet—and of following a path along the River Teign when whole vales of bluebells were in bloom, and of waking up in a train in Cornwall and for the first time seeing foxglove growing wild. I shall never forget, either, riding alone through dusty gray sagebrush up a sandy foothill of the Sierras and coming upon a fenced pasture that would have been as green as the grass of Kentucky except that in the grass grew a multitude of wild flowers in bloom. The pasture, as it happened, enclosed a lone and startled rearing stallion, and I was riding a mare, so I turned back fast, not daring to dismount to study the flowers. Thus the sight of that tapestried field has assumed a dreamlike quality, like a glimpse of the garden of the gods on Mount Olympus, with its fields of asphodels.

Last summer, for the first time, the roadsides of our town in Maine were sprayed with herbicides, both by the Telephone Company and by the State. Up till now, they have been a delight in spring, with flowering trees—wild pear and wild cherry—as a background, and right beside the tar the white flowers of wild strawberry, purple violets, blue star grass, and yellow clover (the tiny Lesser Yellow Trefoil), and in midsummer and early autumn with goldenrod, blue asters, and white ironweed. Perhaps all these will be gone another season. I take comfort that the flowers at least will live on forever, in all their lovely colors, in the pages of *Wild Flowers of the United States.*

10

The Flower Arrangers

Amaryllis. From Breck's, 1908

November 4, 1967

Flower arrangement, which today has Americans in its grip—especially American women, myself included—has been going on since long before the birth of Christ. The evidence is there in the scenes on Egyptian vases and in tomb murals, in the patterns on early Persian carpets and brocades, in Greek sculpture, in Roman and Byzantine mosaics, and in Chinese, Indian, and Japanese art of all kinds—lacquers, screen and scroll paintings, porcelains, bronzes, and carved ivories. The progress of the art (or should I say vice?) over the ages is well and briefly described and voluminously illustrated in color and black-and-white in a book titled *Period Flower Arrangement*, by Margaret Fairbanks Marcus (Barrows, 1952). The background and the history are accompanied by suggestions on how to make flower arrangements that are in the manner of one period or another in history. For once, in the flood of books on how to arrange flowers, the modern so-called "period arrangements" shown in the illustrations are almost universally lovely. I would be proud to have accomplished any of them myself, even though I have never had much desire to make a period arrangement. Once in a while, I find I can't help making one inadvertently. Last summer, I picked what annuals had to be picked one morning in order to keep the plants in bloom, tucked them without thought into an Italian pottery jar, and suddenly found I had a multiflower, feathery concoction that looked quite like an Odilon Redon flower painting. (Feathery bouquets are not my habit, and I would rather have turned up with a Cézanne period piece than with a Redon.)

Mrs. Marcus tells us that the earliest record of flowers arranged *in a container* in the Western world is a first-century-A.D. Roman mosaic, which the accompanying illustration shows as a graceful and unpretentious basket of roses, anemones, pinks, and smaller, unidentifiable flowers. An even more interesting illustration is of a fourteenth-century-B.C. wall painting from the tomb of King Apy in Thebes, the capital of ancient Upper Egypt during its greatest period of art and empire. The King, seated with his Queen near a table set with food, holds a formal, torch-shaped bouquet of lotus and lotus buds, their stems bound together tightly and covered by horizontal rows of petals and berries, in red, pale blue, and blue-black, making a formal design. A similar bouquet lies on a table in the background, where a second banquet is set forth; flowers and food were among the traditional Egyptian offerings to the dead. Another figure in the mural—servant or offspring?—appears to be handing the King still a third bouquet, held in a rather clumsy cylindrical holder with a rounded bottom, but I have not seen the mural, which is now in the Metropolitan Museum of Art, and if I had I would not be enough of an Egyptologist to know what this offering is. It *looks* like the tops of flowers, with the stems and leaves completely hidden in the cylinder. The judges of modern flower shows might be gratified to note the absence of visible stems in these ancient bouquets; not a stem shows in any of the Egyptian flower offerings. In fact, the flower arrangements of Thebes were as unflowerlike and as stiffly stylized as many of the exhibits in our contemporary flower shows. (I might explain to those who have never entered an arrangement in a modern flower show that the so-called accredited judges of what are known as "standard" flower shows sponsored by the National Council of State Garden Clubs do not usually approve of glass containers through which the stems of flowers can be seen, especially if the stems cross one another. I don't belong to a garden club, but I have entered a few bowls and vases of flowers in the pleasantly unpretentious flower show put on each summer by the garden club of my small Maine town. Happily, our club doesn't attempt to meet the regulations of a standard show, but sometimes an "accredited" judge is pressed into service, and I learned about this objection to crossed stems from one such visiting authority. I happen to love flowers in clear glass, provided the water is not cloudy or dirty, and since the stems of many delicate flowers—Shirley poppies, for instance—crisscross in my garden beds, I find them not unpleasant or unnatural to look at when they are crisscrossed in a transparent vase.)

To revert to our ancient history, Egyptians liked to carry formal bouquets and to arrange flowers stiffly in basins or in wide-mouthed vases with tapering pedestals or in tall "spout vases," but they were also as famous as the Greeks and the Romans for their wreaths, garlands, and chaplets of flowers. In general, the garland, the tray heaped with flowers, and the cornucopia filled with fruits and blossoms were the most common and delightful arrangements of the classical Mediterranean world. Byzantium, however, the Eastern capital of the Roman Empire, was influenced by early Oriental art, and Byzantine bouquets were more formal than the classical Mediterranean, tending to the spire or cone of green studded with flowers, rather like our modern florists' Christmas-dinner centerpieces of evergreens trimmed with small Christmas-tree ornaments.

From the earliest days, the symbolism of flowers was an important element—probably most of all in the Orient, and especially in China and Japan. Japan, with its emphasis on ritual, has made the biggest fuss about arranging flowers and had the greatest impact on Westerners. Dedicated flower arrangers in this country and Europe must by now be soaked in the history of Japanese *ikebana* (the Japanese word for "flower arrangement"), since every season brings a new set of books on the subject, yet I'm sure most of the Occidental writer-arrangers must still be floundering in the details and nuances of the rules, just as I am. I have made an honest effort to absorb the principles and the strict laws of the Japanese art in most of its branches, but the books have different and often conflicting explications and interpretations, whether written by Japanese or by European, English, or American writers, and, to add to the confusion, every writer in English translates the meaning of the Japanese ideographs slightly differently and spells them wholly differently. I shall sum up my muddled impressions of the history and complexity of the art of *ikebana* as best I can. I do so in part because this Japanese emphasis on rules has certainly had great impact on the flower shows held under the auspices of the ubiquitous National Council of State Garden Clubs. These garden-club rules, which we'll come to eventually, have little to do with the art of *ikebana*, but the Council's idea of strict requirements for putting flowers in vases seems to have stemmed originally from the Orient.

The very earliest Japanese arrangements were offerings of cut greens at Shinto shrines; no flowers were used. The use of flowers and of colors other than green came to Japan from China in the seventh century A.D., with the earliest Buddhist priests, but it was not until the fifteenth century that a number of *ryu*, or schools, of formal flower arrangement began to flourish. There are many, many more there today, specializing in the various styles, such as Rikkwa (or Rikka) and Shoka (or Sheika or Shakei), the classic ancient forms, or the more recent but still early styles, such as Nageire and Moribana, each one of which appears to have myriad variations. The most recent of all is called Free Style, and *it* has three branches, with the revealingly Occidental designations of "avant-garde," "objet," and, inevitably, "abstract." But this free-style method, derived from the West, is quite out of the

Japanese main line of flower arrangement and its complex flower symbolism.

One can get at least a misty idea of the whole elaborate business from two books out of the many scores available—*Japanese Flower Arrangement: Classical and Modern*, by Norman J. Sparnon (Tuttle, 1960), a large volume illustrated with a hundred not very good color plates and even more numerous photographs, drawings, and diagrams, and *The Masters' Book of Ikebana*, which was published last year. The latter, a huge volume weighing a ton and by far the more beautiful book of the two, though not necessarily the more reliable, is bound in green Japanese silk brocade and is a publication of Bijutsu Shuppan-sha, Tokyo. The book is edited by Donald Richie and Meredith Weatherby and contains a useful chronology and some long historical-background chapters by Mr. Richie, followed by lessons from the headmasters of three of today's most famous schools of flower arranging—the Ikenobo, the Ohara, and the Sogetsu. The illustrations—more than four hundred of them, a hundred and twenty of which are in color—are what really matter most to me in this de luxe production. They start out with the delightful *ikebana* to be seen in old Japanese prints and floral manuals and in ancient paintings on scrolls, screens, and wooden panels, and progress to photographs of the contemporary Japanese arrangements that illustrate the lessons by the modern masters. The jump from the ancient past to the present is rather shocking, for the photographs of the early examples are indeed visions of beauty—as art, as arrangements of flowers, and as color reproductions—whereas the modern *ikebana* (to these eyes, anyway) are quite often hideous both as arrangements of flowers and as color photographs, and sometimes they come near to being ludicrous. Some do retain the old charm. Yet if this Japanese national preoccupation seems to me to be on the decline as an art, it is surging forward as a business. The Ikenobo School alone, which claims to have been founded in 1462, has now two institutes and three hundred branch schools and three million students in Japan itself, and there are branches of the school scattered over the globe, five of them in this country. As for the text of the book, Mr. Richie's historical chapters, which make up the first half, attempt to cover, in orderly progression and in all too Occidental terminology, the periods and the varying habits and styles of the Nipponese way with flowers, from prehistory to the present. As for facts, he adds some useful new material, but for an interpretation of the spiritual meaning and symbolism of it all I find his chapters less illuminating than the books by Gustie Herrigel and Lafcadio Hearn, which I shall describe later.

As I've said, no two people agree on the history and techniques of Japanese flower arrangement, but I shall go out on a limb and continue to give my own ideas of the basic principles of *ikebana*, drawn from many books and sources. For the traditional classical styles, the student must first learn "The Principle of Three." This means three branches in a container (or it could be one stem with three branches), the tallest of which is called *shin*, for Heaven, the medium length *so*, for Man, and the shortest *gyo*, for Earth. This is the triad that, to the Japanese, represents the cosmos. As for symbolism, most of which came to Japan from China, there is far more to it than the triad. Not only is every flower or tree a symbol of something—a plum blossom signifies "resistance to injury" and also "new hope"—but there is a more general symbolism, which is the Chinese principle of opposites called Yin and Yang, or, by the Japanese, Yin and Yan or In and Yo. (I warned you!) Yin, according to Webster, is the feminine, or negative, and dark principle; Yang is the male, or positive—the bright and beneficent principle. Theoretically, they should both be embodied in every Japanese arrangement. (One might note here that for centuries men were the Japanese flower arrangers and that only in modern times have women been allowed to study the art.) Gustie Herrigel, the author of *Zen in the Art of Flower Arrangement*, however, puts it in a way more kind to my sex. She says that Yo-Yin (as *she* terms the opposites) is meant to demonstrate not a contrast but a principle of balance. "Yo," she writes, "signifies the sunny side . . . [and] symbolizes the generative, flowering principle of Nature. Its colours are red, purple, pink. Yin, on the other hand, is receptive. It is thought of as dark, lying in the shadow. Its form . . . is suggested by buds and curves. It is correlated with the left side, and its colours are white, yellow, blue." Correlated with the left side? Yes, there are right-sided and left-sided Japanese bouquets, each with its meaning, but here, dazed, I stop short in my summary, lest it become "confusion worse confounded." It is safe to say that in the Orient it was, and is, Yin and Yang, Yin and Yan, In and Yo, and, one might add, Zen and all zat.

I mean no irreverence to Zen Buddhism, which I hold in respect; by all odds my favorite modern treatise on Japanese flower arrangement is the book I have just quoted from, *Zen in the Art of Flower Arrangement*. It is a small and, to me, moving book, translated from the German by R. F. C. Hull, and first published here in 1958 by the Charles T. Branford Company. I commend it to your attention. The brief foreword, by Daisetz T. Suzuki, gives the perfect contemporary explanation of "the breath of the spirit" that binds art and religion together in the history of Japanese culture. "The art of flower arrangement is not, in its truest sense, an art," writes Dr. Suzuki, "but rather the expression of a much deeper experience of life." Most Occidentals who try to express or imitate this deeper experience have not had much luck, if one is to judge by the photographs of American arrangements in the Japanese manner that pepper the current books by English and American writers on the Japanese way with flowers. There are exceptions, but for the most part our Occidental attempts at Oriental simplicity, style, spirituality, and symbolism seem to miss the mark. We can learn the techniques and copy the styles, but Oriental symbolism transplanted to the Western world is nearly always a bit ersatz. Our sources of inspiration are inherited from the West and are quite different, and our approach to the natural world and to flowers, while often fervent and even religious, is a different approach. A few final quotes from Mrs. Herrigel might help prove my point:

From whatever side the Westerner seeks access to the spiritual life of the East, he will encounter quite special difficulties. Almost always he is in danger of wanting to penetrate intellectually into what lies beyond the intellect, into something that is given to Eastern man directly, and which he experiences in unquestioning reality. . . .

Even though there are many things in flower-setting that can be said and shown, yet behind everything that can be visibly represented there stands, waiting to be experienced by everyone, the mystery and deep ground of existence. . . . Not the slightest intention of arranging [flowers] "beautifully" must disturb this self-immersion.

She also explains what most Westerners now know—that a Japanese garden is not a place to grow flowers and that in Japan there are very few meadows alive with wild flowers. The Japanese, young and old, do not pick wild flowers. They revere them in their natural setting and they go to view special beauty spots and flower displays. The flowers they arrange for their homes are usually grown commercially. As everyone knows, most Japanese homes have a niche, or *tokonoma*, where arrangements are placed in a spirit of dedication, to be viewed for their spiritual value by members of the family and their visitors.

I and doubtless thousands of other Americans who are old enough to remember the late eighteen-nineties and early nineteen-hundreds got our first knowledge of Japan, Japanese gardens, and the Japanese attitude toward flowers from Lafcadio Hearn's *Glimpses of Unfamiliar Japan*, which was published in the United States in 1894. My elders had read it in a public-library copy, but the book was given to me at an early age—perhaps I was twelve—by my New England aunt who had taken the then astounding step of marrying a Japanese. I found it delightful, and for years I cherished its two volumes and sometimes reread them. Hearn was born on one of the Ionian Islands, in 1850, of an Irish Catholic father and a Greek mother, and his Japanese-sounding first name was originally spelled "Lefcadio," a name his parents based on the name of his island of birth, Leucadia—which, says the Encyclopaedia Britannica, was pronounced "Lefcadia." Later, Hearn himself changed the spelling to "Lafcadio." At the age of nineteen, he came to this country, and he lived on a meagre income from sporadic journalism until, in 1887, a New Orleans paper sent him to the West Indies. In 1890, *Harper's New Monthly* sent him to Japan as a foreign correspondent. The latter assignment did not last long, but in Japan Hearn found his spiritual home and a post as teacher of English at the University of Tokyo. He married a Japanese, became a Buddhist, and was naturalized under the name of Yakumo Koizumi. *Glimpses* was his most famous book, but eleven more about Japan were to follow before his death, in 1904. His love of the ancient and as yet un-Westernized Japan and his detestation of Occidental ways with gardens and cut flowers run throughout this book. Hear him in Volume II of *Glimpses*:

After having learned—merely by seeing, for the practical knowledge of the art requires years of study and experience, besides a natural, instinctive sense of beauty—something about the Japanese manner of arranging flowers, one can thereafter consider European ideas of floral decoration only as vulgarities. This observation is not the result of any hasty enthusiasm, but a conviction settled by long residence in the interior. I have come to understand the unspeakable loveliness of a solitary spray of blossoms arranged as only a Japanese expert knows how to arrange it—not by simply poking the spray into a vase, but by perhaps one whole hour's labor of trimming and posing and daintiest manipulation—and therefore I cannot think now of what we Occidentals call a "bouquet" as anything but a vulgar murdering of flowers, an outrage upon the color sense, a brutality, an abomination.

Earlier, in Volume I, when he is describing a *matsuri*, or sacred festival, held at night in the streets of a small city, he comes upon a silent crowd gathered around a street booth:

As soon as one can get a chance to look one finds there is nothing to look at but a few vases containing sprays of flowers, or perhaps some light gracious branches freshly cut from a blossoming tree. It is simply a little flower show, or, more correctly, a free exhibition of master skill in the arrangement of flowers. For the Japanese do not brutally chop off flower heads to work them up into meaningless masses of color, as we barbarians do: they love nature too well for that; they know how much the natural charm of the flower depends upon its setting and mounting, its relation to leaf and stem, and they select a single graceful branch or spray just as nature made it. At first you will not, as a Western stranger, comprehend such an exhibition at all: You are yet a savage in such matters compared with the commonest coolies about you. But even while you are still wondering at popular interest in this simple little show, the charm of it will begin to grow upon you, will become a revelation to you; and, despite your Occidental idea of self-superiority, you will feel humbled by the discovery that all flower displays you have ever seen abroad were only monstrosities in comparison with the natural beauty of those few simple sprays.

Lafcadio Hearn was a reporter with an eye for beauty, and he could write, but, illuminating as his key book was and is, I feel that it may have presaged and hastened the present-day self-consciousness of the earnest American flower arranger. Hearn could hardly have realized what a floodgate he had opened, and in this book he gave no hint of knowing the strict rules that presumably governed the hours of patient, loving labor that went into the placing of his solitary spray. Better than most Westerners, though, he grasped the spirit and dedication behind the Japanese arrangers' "master skill."

Glimpses of Unfamiliar Japan had its good results and its bad. Certainly, in our New England household, some of them were good. The odd-shaped Tiffany "greenery-yallery" glass bowl, filled, usually, with fat pink store carnations, which used to stand anachronistically on my Maine grandmother's small Chippendale sewing table between the French doors of our pink parlor, was eventually banished to the dining room, where it was used for party centerpieces on the round mahogany dining table. There, at Christmas, it was filled with scarlet carnations and asparagus fern or with red Hadley roses, but for dinner parties throughout the year it always held the same oversized pink carnations, to match the pink shades clamped to the candles in our silver candelabras. Our parlor, meanwhile, had been transformed. A dim gold tea paper (or perhaps it was an approximation thereof) replaced the pink-striped wallpaper my father had inherited when he bought the house, the ruffled point-d'esprit glass curtains on the French doors were removed in favor of long, soft-green brocade draw curtains, and for the little sewing table between the doors my father purchased, at Yamanaka's, in Boston, a tall, pear-shaped, pale-gray ceramic vase with a dragon motif. In this vase we arranged, without much study or prayer, to make a Japanesy effect, a few long branches of flowering shrubs from our borders—forsythia first, of course, followed by one spreading double or triple branch of Japanese crab apple, with its delicious rosy-red buds and pink flowers. Next came the bold coral of Japanese quince, then English hawthorn (pink, white, or red), which grew so tall that my next-older sister and I had to gather it by scrambling up the high lattice fence that separated our yard and border of shrubbery from the empty lot next door. Lilacs, purple or white, were gathered in the same way, also because the bushes were so tall. By lilac time, the single Japanese branch had long since been forgotten and the dragon vase became a mass of color, or what Hearn would have called a monstrosity and a vulgar murdering of flowers. In June, it held peonies or Bridal Wreath spiraea, again en masse, and by autumn we had reverted entirely to Victorian ways and filled the vase on football weekends with those huge boughten chrysanthemums that, I'm sorry to say, are now called "football mums." We had them in a succession of colors—white, yellow, bronze, and, for the day of the Harvard-Yale game, a good Harvard crimson. I thought they were gorgeous.

This is perhaps the place to recall some of the other flower containers of the first decade of this century; in a way, they controlled the bouquets of my parents' generation and, handed down, control mine. I inherited quite a number of them, and very likely they will someday be used by my children and grandchildren. Every American family in that now distant past had its rose bowl—a round glass vase with a fairly small opening. Some were plain, more were ornate. We had two—a big thin clear-glass affair, like a goldfish bowl, and a heavier small pressed-glass bowl with a diamond pattern, very pretty for rosebuds. My husband's mother's large rose bowl was more elaborate, and I cherish it, though I seldom use it for fear of hurting it. It is medium-sized, and it, too, is made of patterned pressed glass, but it is also decorated with a delicate gold-filigree pattern of grape leaves and grapes. Set into the gold design near the top are four tiny ruby-red glass gems. The mouth is wide enough to hold six or eight roses, and I can imagine that it must have been used for many a delightful Victorian bouquet of dark-red roses, which might have been the fragrant Général Jacqueminot or Prince Camille de Rohan, two of the great garden reds of that era.

Though for the most part I've abandoned rose bowls, I do constantly use a whole series of hammered copper pots or jars of varying shapes and sizes, which we had in our house in Brookline from as far back as I can remember. Well polished, they are perfect containers for the more sturdy garden flowers. The huge one I fill with decorative sunflowers that range from yellow and gold to henna and maroon, or with Mr. Burpee's very largest Climax African marigolds—Toreador, Primrose, and the others. The next-biggest jar is good for zinnias, chrysanthemums, and mixed flowers, and the smallest is just right for calendulas. In my youth, our Paul Revere silver tankard was considered too precious to hold flowers, but nowadays, from June to November, I keep it filled with a few roses in the open-shelved dining-room corner cupboard that holds what remains of the ancestral Lowestoft.

A friend has reminded me of the very tall vases, shaped much like umbrella stands, that were especially designed to hold the long-stemmed fragrant American Beauty roses when they were the rage. We owned no such vase because we seldom indulged in American Beauties. My father considered them a bit "gross," with their great size and strong blue-pink coloring, but this may have been an aesthetic verdict influenced by the pocketbook; the Beauties were very costly for their day. I remember the tall vases, though, in the houses of friends where the American Beauty happened to be the most prized rose. It was indeed so much in vogue that one of the bolder débutantes of my five-years-older sister's generation had a costume made by her dressmaker for her year of "coming-out" balls that was designed to be worn with two choice American Beauties. The floor-length gown, as I recall it, was made of rich white satin with a very low neck, and with it she wore, of all things, bright-green silk stockings. In the V of her low neckline she tucked two huge American Beauty roses. If one of her partners commented on their fragrance or their rich color, this saucy girl was reputed to lift her floor-length skirt daintily to show her pretty green ankles and maybe even a bit of her slim green calves, and say demurely, "Yes, lovely. And here are the stems."

Despite these remembered excesses, I can't agree with Hearn about the total vulgarity of the Western world's tradition of flower arrangement, and I can only guess that his book had an influence on our transformed parlor. After all, the taste of my elders was changing, as everyone's does, with the times. To my eye, however, some of the most unfortunate treatments of flowers one sees in today's flower shows and books on flower arrangement are the overstudied, imitative approximations of the Japanese simplicity and stylization, put together without emotion, meaning, or understanding of the spiritual fervor behind the style that is being copied. The only modern trend I dislike more is the quite recent attempt at flower abstracts, which at the flower shows often tend to make me laugh, though by no means always. I have no prejudice against abstract art per se or against Japanese arrangements by Occidentals, which can be a complete delight. A pure abstract montage made of blossoms might also be a fascinating if perishable objet d'art. (I have yet to encounter Pop Art imitated by arrangers. However, a scarlet geranium grown in a Campbell's tomato-soup can, as I've seen it grown in country kitchens long before the days of Andy Warhol, would be better Pop Art for me than flower arrangements imitating Warhol and his colleagues could ever be.) In the hands of an artist with flowers, almost every style can be a lovely thing to look at, and even a work of art (or at any rate sub-art). I think, though, that Americans are most successful when they do not copy any style, or at least keep to their own natural inheritances from the Western world—the English, the French, the Italian, the Flemish, the German, and the Austrian.

For example, Jan Brueghel and his followers were painting big armfuls of assorted flowers on canvas as early as the late seventeen-hundreds. "Blumenstrauss mit Schnecke" ("Bouquet of Flowers with a Snail"), now in the Strasbourg Musée des Beaux-Arts, which shows a great gathering of mixed blooms, such as tulips, iris, lilies, poppies, and smaller flowers, in a fat wooden tub (a predatory snail at its base), is the sort of flower arrangement any American not too bound by modern trends could and would turn out today provided she had a large garden to pick in. So are the more florid but luscious bouquets, carefully arranged and realistically reproduced on canvas by the better flower painters of the Biedermeier School, such as Johann Knapp and Sebastian Wegmayr. Victorian England may have been a bad influence except for those who followed the revolution against patterned garden beds and Victorian fussiness in general started by Gertrude Jekyll, who was an arranger as well as a gardener. On the whole, though, English flower arrangers of all periods have been among the best, and so have their books.

From Joan Parry Dutton, an Englishwoman turned American, we get to know both English eighteenth-century and Colonial Virginia's habits with gardens and bouquets in her small book, *The Flower World of Williamsburg*, a publication of Colonial Williamsburg, distributed by Holt, Rinehart & Winston (1962). The book is a more serious study than its title indicates. The early chapters—on the formal gardens of the eighteenth century in England, on Virginia's simpler but imitative gardens when it was a royal colony, and on the first American plant explorers and their interchange of plants between the mother country and the colonies (and, later, the botanists of the young Republic)—are scholarly, lucid, and full of odd and stimulating bits of information. One item I like is that while prosperous Virginia planters were filling their gardens with importations from England, the gardeners of the great English estates were busily planting long borders of an American weed called goldenrod. Mrs. Dutton introduces us to Queen Mary II, a woman after my own heart, who is known historically, the author says, as England's "first flower arranger." Her influence in early Virginia was strong. When she and her King, William of Orange, took over Hampton Court, he devoted himself to rehabilitating the famous formal gardens, but it was Mary who insisted on adding flowers to relieve the severity of the green topiary trees and clipped hedges. She also brought flowers galore into the palace and she demanded and got good ceramics (some especially designed for her) to hold her bouquets, most of them blue-and-white delftware from Holland—pots, bowls, jars, and "posy-holders." These last are partially covered containers with holes in the top for the insertion of flower stems, but aside from these special vases and a queer invention of her own—a flower table with a perforated top—the young Queen had no holders or props for her blooms. (Today's Williamsburg craftsmen have copied many of these old delft vases; some of them are both pleasing and useful.) So far as is known, there were no rules for bouquet-making either in Hampton Court or in Williamsburg in its great days—something we would do well to remember today.

Another of our more fortunate inheritances as flower arrangers, were we but to take advantage of it, comes from modern painting—or, rather, from the great modern artists in all mediums. I have been in the homes and studios of a few of them in this country and have always seemed to see flowers there but never flower arrangements of the studied sort one finds in the "artistic divisions" of the big flower shows, which appear to be singularly untouched by any modern Occidental art except the most avant-garde. There are many photographs that show what I mean in that revealing book by Alexander Liberman, *The Artist in His Studio* (Viking, 1960), in which the author-photographer explores "the mystery of environment" and captures brilliantly the ambiance of the great modern French painters, living or dead. He starts with Cézanne, Renoir, and the other Impressionists and Post-Impressionists, and covers all the important schools down to the contemporary French sculptors and painters—contemporary as of the late nineteen-fifties. Flowers apparently played a major role in the lives of a good many of them, and were, indeed, a part of their environment. Most of the artists Liberman documented in text and photographs were alive; luckily, artists are long-lived, and, even more luckily, the studios and sometimes the homes of some of the earlier moderns, long since gone, had been kept much as they were when the great men were alive. Cézanne's studio, "the birthplace of modern art," was still intact in 1949, when Liberman photographed it, and Monet's Impressionist garden, with its blocks and drifts of colored bloom and its secret Oriental lily pond, was still in flower at Giverny when the author visited and photographed it in 1954. The Monet house was then occupied by his stepson, who had continued to fill it with vases of the flowers that meant so much to Monet and his art. Liberman visited Braque's Norman peasant house and vast studio in Varengeville when the painter was seventy-eight. In the studio, Liberman writes, there were, wherever he looked, small bouquets of bright-red flowers that corresponded with the bright-red spots on Braque's otherwise monochromatic gray palette. A big ochre pitcher filled with tawny single dahlias was photographed in the Braque house. It was arranged by Mme Braque, and it is decorative and artful in a good way, but it is perhaps the most "arranged" vaseful of flowers shown in the volume. Most of the bouquets are like the handful of field daisies in a jug—a bouquet a child might have gathered—in the photograph of Giacometti's bare, shabby bedroom. In the excellent color plates of the modern French paintings, of which there is a representative selection for each artist, there are to be seen on the canvases other jars, jugs, and vases of flowers, not merely in the still lifes of flowers but in the larger compositions, in which they play a more incidental part. Whatever the period of a painting, not a vaseful shown in the volume looks consciously set up to make a decorative or modish effect. As in the homes and studios, in the paintings they seem to be there because the artist or a member of his household liked flowers enough to bring them indoors. Naturally, in the paintings, since they *are* part of a composition, they are to that extent "arranged," but I doubt that the judges of a modern flower show, with their book of rules, would give them a second glance. For instance, in Matisse's "The Hindu Pose," perhaps the best known of his famous "Odalisque" series, there are some unrecognizable lavender flowers (they just *might* be wild flag)—five long stalks of even length, seemingly thrust without apparent plan or style into a two-handled green pottery jar, which stands on a table beside the semi-nude model, who sits, cross-legged, on a blue-checkered chair or divan. Obviously, the flowers are there for their color, mass, and balance in the whole composition and not because Matisse wanted to paint a carefully arranged vase of flowers.

This seemingly artless use of flowers in the home—or in art—to me is one good part of our Western inheritance, and better for Americans to aim at than imitating a complicated Oriental floral tradition that is not ours or trying to create a modern abstract out of dead vines or broken cattails just to be in style. For one thing, we Americans are too impatient a people to create Japanese arrangements even if we grasp their meaning to the Japanese and learn to feel, as Lafcadio Hearn did, the spirit behind the arrangement of the flowers. As for today's avant-garde bouquets, they seem to lose sight of flowers entirely. They may be fun to do as a sort of game, but they are often not even "living sculpture," as the modern Japanese teachers like to call their arrangements. ("Living," by the way, is a wrong term for any cut flower; one should never forget that anything that grows, once cut, is dying, unless it is a vine or leaf being rooted in water or sand.) This is why I prefer my modern sculpture straight and in a substance that endures.

Our complex Occidental background, with its many, varied artistic influences, is important, but I like it better when Americans are simply themselves—unpretentious gardeners, farmers' wives, city dwellers, or suburbanites—bringing in a few flowers to brighten up the house and arranging them in any way that happens to please them, considering no inheritance or style, following no rules, imitating no other civilization, and using the nearest container that comes handy, whether it be an old beanpot or pickle jar, an expensive Steuben or Finnish crystal vase, a piece of antique china or a modern ceramic.

Best of all is when a modern flower arranger develops his or her own style. I cite as an example my friend Doris

Wolfers, whose living room in summer is always bright with many small vases of flowers set about on low tables. The blossoms are in flat bowls or barely rise above the level of the tops of low vases and are crowded together to form a color pattern, like a jewelled inlay or a richly hued mosaic. They are the blossoms of the annuals she grows in her Maine garden, usually arranged without foliage so that only the bright purples, violets, reds, and golds of some such flower as salpiglossis make the pattern. There they gleam as one looks down on them—a flat composition made of bloom, glowing with the colors of old cathedral stained-glass windows on a sunny day. We have all of us, I guess, made flat, patterned arrangements by floating pansies or violas in a shallow dish, but to lop off the heads of taller flowers like salpiglossis, verbenas, annual phlox, and all the others and arrange them in a pattern of blossoms is unique in my experience. I asked Mrs. Wolfers how she came to develop this style, and she told me that when she was a girl in Switzerland and went on skiing or climbing parties in the Alps she and her friends would gather the Alpine wild flowers on the lower slopes of the mountains and bring them back to the hotel where they were lodged. Usually the only containers available were hotel soup plates, so she arranged the heads of her Alpine flowers to make an interesting pattern in these shallow dishes. From this beginning, Mrs. Wolfers, who is better known for her embroidered montages, developed a style of flower arrangement that is just right for an artist whose medium emphasizes texture and color.

Where did this word "arranged" come from, to start with? In my girlhood, the question at breakfast was "Who will pick the flowers today?" (This meant who will pick *and arrange* the flowers.) If I had been English, it would have been "Who will do the flowers?" The word "arrangement" is newer in this country than we realize. Many garden flowers have to be picked if they are to continue to bloom, and houses from the earliest days in Europe and America were brightened by color from outdoors. Where there are small children (except, possibly, in the Orient), flowers will be instinctively gathered for their brightness alone. What country mother has not helped her child tuck his gathered wilting dandelions, daisies, and buttercups into a jelly jar or tumbler in order to preserve in water, if only for a day, these bright "objects of delight"? I daresay the earliest cave woman brought flowers into the cave.

Flowers have been a traditional offering to the gods (and also to the beloved, alive or dead) in many very different civilizations. And artists from the earliest days of Western art have painted them. Growing flowers and flowers in vases and bowls appear in Christian religious art even as far back as the Gothic period, and they form backgrounds for the Mother and Child during the great ages of Italian and Flemish religious art. (The chapter on the Renaissance in *Period Flower Arrangement* shows many such paintings; those from Botticelli and his school are outstanding for their use of flowers.)

Anyone who has read Kilvert's *Diary* in an uncut version will remember, from his far later period, his long, ecstatic entries for Easter Eve and Easter Day, April 16 and 17, 1870, which describe, more hauntingly than anything else I have ever read, the English and Welsh use of flowers for a religious purpose. The entries are far too long to quote entire, and to condense them at all is to lose much of their essence. The William Plomer one-volume selection from the long *Diary*, with its excellent preface by A. L. Rowse (Macmillan, 1947), reduced them to a brief extract that catches little of the quality of the whole. Therefore, if you can't find Plomer's three-volume uncut edition, published by Jonathan Cape in 1960, I suggest a very fine recent anthology of English nature writings, *The Poetry of Earth*, selected by E. D. H. Johnson (Atheneum, 1966), in which the "Easter Idyll" is reprinted complete except for one paragraph. When this section of the *Diary* was written, Francis Kilvert was a young curate in the Welsh village of Clyro. He was, it turns out, a flower arranger, and many of his parishioners came to be so, too, under his direction. His descriptions of the dressing of the graves in the church graveyard on Easter Eve—a custom not unlike our cemetery offerings on Memorial Day—and his account of his Easter Day that year of 1870 are classic examples of the Occidental natural feeling for flowers and their use in a Christian religious context.

Condensing Francis Kilvert is painful, and I quote from these touching entries at some length, though still cut too much, if only to show what I mean, and in the dim hope that our dressing of graves on Memorial Day (and especially our increasing trend to use artificial flowers for this purpose) and our often garish and stiff store-bought funeral "sprays" and "sheaves" made of real flowers might be forgotten, so that we could return to the combination of love, sorrow, joy, memory, and religious feeling that Francis Kilvert and his villagers achieved by their mossy crosses and artless bunches of freshly gathered wild primroses:

Saturday, Easter Eve, 16 April, 1870

I awoke at 4:30 and there was a glorious sight in the sky, one of the grand spectacles of the Universe. . . . I got up soon after 5 and set to work on my Easter sermon getting two hours for writing before breakfast. . . .

At 11 I went to the school to see if the children were gathering flowers and found they were out in the fields and woods collecting moss, leaving the primroses to be gathered later in the day to give them a better chance of keeping fresh. Next I went to Cae Mawr. Mrs. Morrell had been very busy all the morning preparing decorations for the Font, a round dish full of flowers in water and just big enough to fit into the Font and upon this large dish a pot filled and covered with flowers all wild, primroses, violets, wood anemones, wood sorrel, periwinkles, oxlips and the first blue bells, rising in a gentle pyramid, ferns and larch sprays drooping over the brim, a wreath of simple ivy to go round the stem of the Font, and a bed of moss to encircle the foot of the Font in a narrow band pointed at the corners and angles of the stone with knots of primroses. . . .

Coming back it was cooler for the fierceness of the sun was tempered and I met a refreshing cool breeze. . . . But now the customary beautiful Easter Eve Idyll had fairly begun and people kept arriving from all parts with flowers to dress the graves. Children were coming from the town and from neighbouring villages with baskets of flowers and knives to cut holes in the turf. The roads were lively with people coming and going and the churchyard a busy scene with women and children and a few men moving about among the tombstones and kneeling down beside the green mounds flowering the graves. . . .

More and more people kept coming into the churchyard as they finished their day's work. The sun went down in glory behind the dingle, but still the work of love went on through the twilight and into the dusk until the moon rose full and splendid. The figures continued to move about among the graves and to bend over the green mounds in the calm clear moonlight and warm air of the balmy evening. Water was in great request for the ground was very hard and dry and wanted softening before flowers could be bedded in the turf. The flowers most used were primroses, daffodils, currant [word illegible], laurel and box. . . .

At 8 o'clock there was a gathering of the Choir in the Church to practise the two anthems for to-morrow, and the young people came flocking in from the graves where they had been at work or watching others working, or talking to their friends, for the Churchyard on Easter Eve is a place where a great many people meet. . . . The anthems went very nicely and sounded especially well from the chancel. The moonlight came streaming in broadly through the chancel windows. When the choir had gone and the lights were out and the church quiet again . . . the schoolmaster and his friend stood with me at the Church door in the moonlight. . . . As I walked down the Churchyard alone the decked graves had a strange effect in the moonlight and looked as if the people had laid down to sleep for the night out of doors, ready dressed to rise early on Easter morning. . . .

Easter Day, 17 April, 1870

The happiest, brightest, most beautiful Easter I have ever spent. I woke early and looked out. As I had hoped the day was cloudless, a glorious morning. My first thought was "Christ is Risen." It is not well to lie in bed on Easter morning, indeed it is thought very unlucky. I got up between five and six and was out soon after six. . . .

The village lay quiet and peaceful in the morning sunshine, but by the time I came back from primrosing there was some little stir and people were beginning to open their doors and look out into the fresh fragrant splendid morning. Hannah Whitney's door was open. I tied my primroses up in five bunches, borrowed an old knife etc. and a can of water from Mary and went to Mrs. Powell's for the primroses Annie had tied for me last night. . . .

It was now 8 o'clock and Mrs. Evans was down and just ready to set about finishing the moss crosses. She and Mary Jane went out to gather fresh primroses in the Castle Clump as last night's were rather withered. The moss had greatly improved and freshened into green during the night's soaking and when the crosses were pointed each with five small bunches of primroses they looked very nice and pretty because so very simple. . . .

There was a very large congregation at morning church the largest I have seen for some time, attracted by Easter and the splendour of the day, for they have here an immense reverence for Easter Sunday. . . . After morning service I took Mr. V.

[Mr. Venable, the vicar] round the churchyard and showed him the crosses on his mother's, wife's, and brother's graves. He was quite taken by surprise and very much gratified. I am glad to see that our primrose crosses seem to be having some effect for I think I notice this Easter some attempt to copy them and an advance towards the form of the cross in some of the decorations of the graves. I wish we could get the people to adopt some little design in the disposition of the flowers upon the graves instead of sticking sprigs into the turf aimlessly anywhere, anyhow and with no meaning at all. But one does not like to interfere too much with their artless, natural way of showing their respect and love for the dead. I am thankful to find this beautiful custom on the increase, and observed more and more every year.

It is hard, for believer or non-believer, to accept Lafcadio Hearn's term "barbarians" after reading the Easter entries.

If Hearn, in the nineties, thought of all Occidental flower arrangements as vulgarities, one can't help wanting to know what he would make of the new trends in the West *after* it had been strongly influenced by the Japanese art of *ikebana.* I began wondering when a sharp-eyed reader of this magazine sent in a photostat of a modernistic arrangement by an American professional flower arranger and teacher named Bob Thomas. It was taken from his fifty-two-page booklet titled "A Concise Study of New Trends in Flower Arranging," which was privately printed in Nashville, Tennessee, in 1965. The arrangement, one of sixty-six shown in gray photographs, is titled "The Road (Subjective-Abstract)," and I'll let Mr. Thomas describe it by his own caption for the illustration:

A free-form earthy-toned container here represents the good earth, in color and in form. A looped piece of Wisteria represents the line beauty created in our expanding highway systems across our country. The one large muted colored Mum is the guiding supervision of the State and National Highway Depts. The green Philodendron Leaf represents the masses of workmen and machinery used in the overall construction.

Our reader pointed out that the only thing missing was a scrap of paper to represent the litterbug.

My curiosity about Bob Thomas was enough aroused for me to try to buy the "Concise Study." I failed, so I don't know its price or whether it's even for sale. Finally, a review copy arrived, with the compliments of the author. From it I learned that Mr. Thomas conducts classes in his Nashville studio and travels about the South by station wagon each autumn giving lessons. He is, the "Study" says, a charter member and past-president of the Nashville Men's Garden Club, a member of the Nashville Chapter of Ikebana International, and past member of the Board of District II, Tennessee Federated Garden Clubs. Incidentally, he has now taught me what a flower arranger's "assemblage" is (a term used by the National Council of State Garden Clubs), since he photographs and describes one of his own—obviously a huge affair suitable only for exhibition halls. Again let the author describe it in his own words:

Forms in Space (Assemblage), two copper drain pipes have been anchored together and placed on a long black base. A barrel hoop is painted orange and placed in interlocked position. A cup-needle holder is painted black to hold the plant material which consists of one umbrella palm painted orange, and a cluster of purple Sea-Foam Static [*sic*]. The collection of things are assembled in orderly fashion giving a feeling of an Abstract flower arrangement.

So now, it seems, Occidental flower arrangers are merely *giving the feeling* of flower arrangements.

Another of my favorite Thomas productions is "Sea Forms," designated as a "Non-Objective Abstract." It consists of a dried Tampa Bay starfish, two red feather dusters, and a four-foot curled piece of a leafless stem of trumpet vine extending upward into space. All the components are stuck into a modern Japanese pottery container made to look like a fishbone. Still another "Subjective Abstract," titled "Space Platform No. 1," needs to be seen to be believed. A black cloud is suggested by a mess of intertwined bittersweet vine sprayed with black paint, and, rising above it, three dried Allium blossoms give "the illusion of being unknown elements of space."

Is this flower arranging or is it a new kind of game entirely? Not all the moderns are as far out in space as Mr. Thomas, but just what *is* happening to the pleasant habit of bringing flowers into the house? The answer changes from year to year, it seems, but there are some signposts and warnings that may lead us a bit farther down this tangled path.

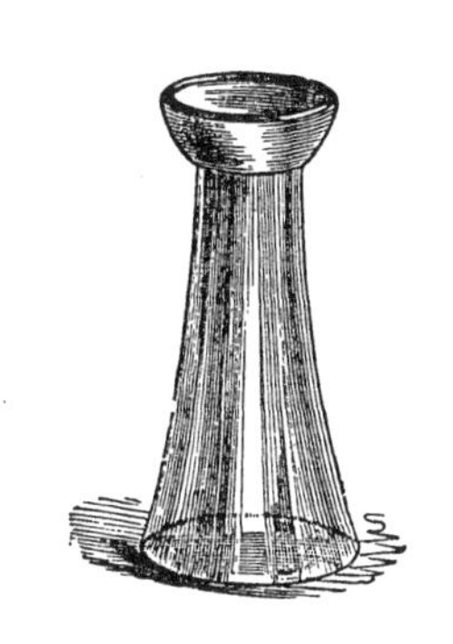

II

More about the Arrangers

Mme Gabriel Luizet roses. From "Harris' Rural Annual," 1897

November 11, 1967

People have been arranging flowers in one way or another since the beginning of civilization. Having wandered hit or miss over the ages, following the course of this odd yet natural preoccupation, I'd like to look at what is going on in the art, or sub-art, of flower arrangement in this country today. The answer is plenty. And the key words are "Garden Club" and "Flower Show." While we do not have several million citizens of both sexes attending schools of flower arrangement, as Japan has today, the number of American amateurs, mostly women, who are happy victims of this obsession is formidable. Of our fifty states, forty-nine have state garden clubs, and almost every local garden club in these forty-nine states, whether a big-city club with many branches or a smaller club of the towns and villages, is a member of its state Federated Garden Club. The forty-nine state clubs, plus a fiftieth (the Garden Club of Washington, D.C.), are again federated under a higher power—the National Council of State Garden Clubs, Inc., which is the controlling nerve center of American flower shows and thus of the aesthetics of flower arrangement.

As a person who likes to grow flowers and bring them into the house, I am all for the local clubs, even though I have never joined one. They often do a great deal of good. They plant the public places that need planting, they teach conservation, they befriend wild flowers, they befriend the birds—natural allies of all gardeners. During the year, most of these small clubs import speakers to talk on such matters as horticultural practice, landscaping, flower diseases and their cure, air and water pollution—any and every subject a gardener is interested in. The little clubs, especially in the summer-resort areas, also provide a democratic meeting place for the local residents and those "from away," and sponsor pleasant social gatherings for many disparate but like-minded gardeners. In general, too, garden clubs pursue the admirable if awkwardly named new goal of Beautification. It was the Mock Turtle, you will remember, who took courses in Reeling and Writhing and in "the different branches of Arithmetic—Ambition, Distraction, Uglification, and Derision." Alice protested to the Mock Turtle that she had never heard of "Uglification." Mrs. Lyndon Johnson, I'm glad to say, has not only heard of it but seen it, and she has made a beginning, with the help of the Congress and the garden clubs, the city fathers and the urban-renewal authorities, toward bringing about a change for the better in our cities and towns and along the highways uglified by man. The results on the highways are still in doubt, with Congress reluctant to appropriate enough money and billboard advertisers alert to preserve their signs. In the course of all this, I fear, Mrs. Johnson and the garden-club ladies have had a few scarifying lessons in Derision.

Though the activities of the local garden clubs are generally benign and beneficial, the scene changes a bit when you get up to the state level, where you run into the subtle influences of the state clubs, with which the local clubs are affiliated. The scene changes even more markedly when you advance to the highest level of all—the National Council of State Garden Clubs. Here you first encounter "suggestions," then rules and regulations, then certified schools of flower arrangement and Council-trained, Council-accredited judges of flower shows, who have attended the Council-certified schools. Here, in short, you encounter "the word." They give it to you straight.

Garden-club judges have a powerful influence on contemporary aesthetics because they judge not only the horticultural exhibits (home-grown potted plants, specimen flowers, and vegetables) but the "artistic divisions" of the Council's so-called "standard" shows. (For "artistic division" read "flower-arrangement division.") Conformity stultifies art, and this club system, as I have observed it at work, does call for conformity despite the comforting foreword to "The Handbook for Flower Shows" of the National Council of State Garden Clubs, a two-hundred-and-thirteen-page manual on the staging, exhibiting, and judging of shows. The book, we are assured, is not a mandatory statement of rules but a guide to freedom "within the framework of National Council requirements." The quoted phrase is the joker. The framework is so rigid that, happily, most of the small-town clubs, for sheer lack of exhibitors and organization, cannot achieve a "standard" show, and their members are thus released from the necessity of conforming and can arrange their flowers in any way that pleases them. Often, though, one or two fully accredited judges, who have gone through the stern disciplines of the Council training, are brought in to judge a small-town flower show, whose exhibitors are thus subjected to the criticism of an accredited person and influenced by his verdicts and his dogma. Judges usually write brief comments of suggestion, approval, or rebuke on each arrangement of flowers exhibited. A mere viewer, like me, can and usually does read these comments. One I read last summer astounded me. One of our best local gardeners and arrangers had used the old-fashioned flower Baby's Breath in several of her entries. She won a blue ribbon with one of these, but even so a visiting judge had written this note to her (I quote from memory): "You really should *not* use Baby's Breath. It would never be accepted in a standard show." (There is, by the way, no rule in the "Handbook" that says this.) Now, I'm not wild about Baby's Breath; I remember from the early part of the century too many florists' bouquets of roses that were ruined by being stuffed with it. But it occurred to me that if Odilon Redon could paint this delicate flower into some of his still lifes of flowers there was no reason at all that a Maine woman should be admonished for including it in her arrangements if it was her pleasure to do so. Moreover, she had used it intelligently, in an original and unusual way, and not as filler material that might kill the outlines of other flowers.

Six times a year, the Council issues a bulletin titled *The National Gardener*, to jack up, instruct, and edify its judges and local officers and its members, who may subscribe to it for a mere dollar a year. A 1964 issue of this publication claimed four hundred and eleven thousand members—a figure that has swelled considerably by now. If you add to this Federated Garden Clubs membership the twelve thousand members of the Garden Club of America (a hundred and seventy-four branches but no control of the branches from above), *and* the members of the totally unaffiliated local garden clubs and horticultural societies, *and* the members of the numerous specialized plant societies, who hold shows of their own, *and* the professional arrangers—florists, teachers in schools of arrangement, arrangers of bouquets for museums and historical monuments like Mount Vernon, Williamsburg, and the White House—*and* the many members of Ikebana Societies, specializing in the Japanese style, *and* (most numerous of all the groups) women, like me, who belong to no club or society, attend no school of flower arranging, and are simply average householders—country dwellers or office workers or suburbanites, to whom arranging flowers is an accepted part of their lives—it is easy to believe that half the women of America are putting fresh flowers into vases fairly constantly. Some of the exceptions are those who buy artificial flowers for their homes and let them stay put, growing dustier and dustier and smelling more and more strongly of plastic as the years pass. Many of them do so for lack of time, lack of a garden, lack of money to buy fresh flowers, or lack of the patience needed to grow live plants. Yet some people actually seem to prefer artificial posies, even people who do not lack for soil, money, time, or patience. There *are* a few good replicas, but no matter how good they lack that essential ingredient, the lustre of life. I could hardly believe it when I came upon an artist painting a still life with phony flowers as her model.

To return to the millions who arrange real flowers, the ones who are influencing taste are the ones who have developed what is now called, in the Council's "Handbook," the contemporary "American style . . . combining certain features of both Oriental and Occidental flower arrangement." What *is* this style today? The "Handbook" gives a part of the answer, as we shall see, and so do the flower shows and the flood of illustrated non-books on how to arrange flowers.

Most flower shows in this modern day are built around a theme. In my area in Maine, the show season opened this year on June 8th, with a combined show of the Brewer Garden and Bird Club and the Bangor Garden Club. The Bangor paper reported that the theme was "Rhapsody in June." The "Handbook" tells us that themes for shows are usual but not required; of recent years, however, I have (with one exception) not visited a flower show outside of New York City that did not boast a theme. Themes, I have discovered, give rise to sub-themes. In the Brewer-Bangor show, some of the sub-themes in the "Artistic Design Division" were "Father's Day," "Spring Ecstasy," "All in the Merry Month of June," "Stag Party," "Rainbow," and "June Breezes—a line arrangement of flowering branches suggesting rhythm, using a bottle as container." The show was a standard show, judged by a battery of judges—most, if not all, of them accredited—and held under the guidance of the National Council of State

Garden Clubs. The usual blue, red, and yellow ribbons were awarded in each class, as well as the coveted Tricolor Award and the Award of Distinction and the Special Creativity Award and the other special awards that are given in standard shows.

The judges who handed out these ribbons, since it *was* a standard show, must have come from among four levels of certification set by the Council, the lowliest of which is a Student Judge who has passed three of her five required courses and is perhaps busy gathering "judging points" in her slow advancement toward accreditation. The requirements for the four levels are set forth in infinite detail in the "Handbook." The Student Judge, it says, must complete five flower-show courses approved by the Council with a passing grade of 70, in a school sanctioned by and under instructors approved by the National Council. She may take no more than two of these courses under one instructor and take no more than two courses in any calendar year; she must obtain exhibiting credits (five blue ribbons at standard shows or at shows of one of the "recognized national plant societies") and judging credits in five similar shows. She must do outside reading for every course (the "Handbook" is part of the required reading for each one) and she must pass an exam on her reading with a grade of 75; the reading exam must be taken no sooner than three months or later than eighteen months after the other courses have been passed. If everything goes along without a hitch and the poor woman doesn't come down with flu and miss a course or fail one and have to repeat it, she will have spent nearly three years on her novitiate before she reaches the second level—that of Accredited Amateur Judge. For the next three years she can relax a bit and go around judging. Then her accreditation expires and she must start refresher courses, furnish evidence of having judged six shows, testify to having submitted five exhibits in standard shows and to being a subscriber to *The National Gardener*. If she can meet these tests, her accreditation can be renewed and she can begin another ascent by attending advanced courses and symposiums. This leads to the low-level summit of Senior Judge, at which she will receive lifetime accreditation. Finally, if her cry is "Excelsior!," the peak itself will come into view, and after two more advanced symposiums and more service as judge and exhibitor, *and* provided she doesn't forget to renew her subscription to *The National Gardener*, she will stand on Darien—a Master Judge, though not necessarily silent. And to think that she perhaps began the long climb only because she loved gardening and flowers!

Ten and a half pages of the "Handbook" are devoted to how to set up and run an accredited flower-show school as part of the Council's system. Each state club is urged to set up such schools. (There are two in my lightly populated state.) The curriculum covers flower-show practices, basic horticulture, design in general, design in arrangements and plantings, period design, specialized studies in four varieties of plants, color, point-judging, ethics in judging procedures, and much, much more. Perhaps the most sobering fact is that all instructors must be approved by the National Council, but no hint is given as to what education, experience, and general background are needed for approval. One cheerful sentence announces that criticism of National Council policies or of the "Handbook" jeopardizes the instructor's standing if the criticism is made to his or her students. He or she may complain only to the Council's National Chairman of Flower Show Schools Committee. Control from the top is the watchword. The courses appear to be crowded into three days twice a year, and must therefore be quite superficial compared to the scientific courses in horticulture, open to amateurs, that are given by the botanical gardens and arboretums and by many of the state universities in their extension work. Some of the state or city horticultural societies also offer excellent courses. For an arranger seriously interested in the principles of design, the art schools and schools of design also offer courses.

The "Handbook's" chapter on "Artistic Design" contains an astonishing collection of didactic material, some of it informative, some of it inadequate (in its historical list of French styles in art and in bouquets, it entirely omits French *modern* art), and more of it abracadabra to a simple gardener who merely knows how to raise flowers and enjoys placing them in vases. Take color. A garden-club judge must, and a garden-club member should, I take it, know how to use the color wheel, understand the color triangle, and distinguish between such terms as "hue," "chroma," "analogous colors," "complementary colors," "split-complements," "tint," "tone," and "shade." Or take design. In the "Handbook" it includes balance, dominance, contrast, rhythm, the parabola, line arrangement, massed line, expressive arrangement, and abstract arrangement. We also hear from the Council on the matter of niches, recessed spaces, backgrounds, table-setting, and "assemblages." The section titled "Occidental Flower Arrangement" starts out with the dubious statement that flower arrangement in the West has never been an "intrinsic part of religious ceremonies"—there goes Francis Kilvert's "Easter Idyll," from which I was gratefully quoting in the preceding chapter—and, forgetting the Greek and Egyptian religious offerings to the dead and the traditional decorations of churches and graves in Europe and America, skips blithely through the centuries in three pages to "Contemporary American Flower Arrangement—20th Century": "Not only have Americans learned to create beauty with design and color, but we have developed the ability to make our flowers speak a message." It strikes me that flowers, since they first appeared on a hitherto barren planet, have never had any trouble speaking their own messages. The messages they speak in flower arrangements, and especially in those created for flower shows, are messages sent out by garden-loving men and women, all too many of whom are now struggling to transmit the messages suggested by their flower-show committees. Before I take up these messages, let's first consider the florists who send floral messages every day in the course of their work. They are, of course, a special case—the true professionals, who make a living by flowers.

Many florists are graduates of commercial schools of floristry. All arrange flowers in the natural course of events, and a few are very good at it. In the biggest cities, there are gifted craftsmen who can create artistic bouquets of all sizes and for all purposes; many of them use unusual flowers as well as the old familiars. In the smaller cities, the flower stores turn out for the most part a standardized arrangement set in floral sponge and graded for price, so you can order by number a bouquet to suit your pocketbook. Yet even florists have their fads in arrangements and follow the trends of the year. Six years ago I was sick in a hospital in a city strange to me and my family. My loyal but distant friends rallied round and wired me flowers. These floral tributes arrived from as many as half a dozen stores in the city, but all of them were as like as two peas, for that year the florists of that city had discovered the flat-faced triangle set in a low container. Most of the triangles rose to a point in the exact center; a few of them bore to the left or to the right, and most of them were made up of gladiolas or chrysanthemums, the florists' standbys. They were pretty, and they all gave me pleasure and saved the day for a lonely patient who had the added pleasure of wondering just how this particular stylized form had happened to hit this particular city that season—some teacher in a local school of floristry, perhaps? Fads also hit the amateurs. A show in New York one year had a vast preponderance of peaked, narrow arrangements rising high and thin, so strikingly so that when August and our village flower show rolled around I decided to make a parody arrangement and enter it. I constructed it to look as tall and narrow and ugly as possible. This was easy, because I happened to have a particularly hideous purple gladiola in my vegetable garden that year and a fine white specimen. Two sprays of purple and one of white with a leaf or two, all crowded together and rising high and thin, made an effect that amused me, but nobody recognized the thing as a parody and I deservedly received a solemn comment on my ill-considered prank. In the same way, the illustrations of flower-arrangement books show the current fads and trends, which stem from many sources. I don't often keep such books, but I saved one from 1958 titled *Flower Show Ribbon-Winning Arrangements*, by Mrs. Raymond Russ Stoltz (Scribners). Some are lovely, some to me look silly (as all farfetched interpretations of a theme do), and a few have that peaked, narrow look that was just coming into style when the book was published.

Despite my reservations about the National Council and the heavy hand it lays on American flower arranging, I love to go to flower shows, even standard flower shows conducted under the Council's aegis. A couple of winters back, I went to one held by the flourishing garden club of a medium-sized Florida city. It was staged in the city's biggest auditorium and was called "A World Tour of Gardens." The flower-arrangement section, so big it occupied more than half of a fifty-page catalogue, came under the heading "Artistic Division: Round the World with Flora." Well, as usual in recent years, there were mighty few flora in the arrangements so lovingly and painstakingly devised by the devoted ladies of the many chapters of the club. Backgrounds and what are called in flower-show language "accessories" ruled the day. There were nineteen classes under this heading of "Artistic," with such titles as "Australia—Land of the Boomerang," "France—Pomp and Elegance," "Korea—Country of the Morning Calm," even "Never Never Land." For each class there were strict requirements. Five of the classes allowed only dried or painted material, seven specified fresh. The others called for a mixture of fresh and dried material, but far more dry stuff than green stuff was in evidence, and this in a flower-laden city at the height of its semitropical bloom. No wonder that when my husband accompanied me to an earlier show in the same building he looked around the vast hall and asked, "Where are the flowers?" This year he waited outside—said he'd seen enough dry-ki to last him—while I fought my way in. (The show had been open an hour and my ticket number was three thousand and something.)

Here, reconstructed from almost illegible notes, is a rundown of a few of the entries:

Iceland: Rules requiring an abstract to interpret the feeling of cold. Some fresh plant material must be used. The blue-ribbon winner's fresh material was one white chrysanthemum inserted in a piece of twisted cedar that had been painted white and rested on a heap of broken pale-green glass. (I own a pair of pale-green glass bookends, and I couldn't help wondering whether the exhibitor had taken a hammer to a similar pair of her own.) This exhibit won the Creativity Award.

Never Never Land: Requirements: "Avant-Garde—Some fresh plant material must be used." A paper-covered cardboard cylinder had been painted red and peppered with red sequins stuck on while the paint was wet. Cylinder held some flowers and artificially curled papyrus. "Camp, perhaps," I noted, "but not Avant-Garde." Won a blue ribbon. Another exhibit was fashioned of two long coils of half-inch copper tubing that had been tortured into intertwining circles. Out of each of the four sawed-off ends of the tubing one small magenta-pink rose drooped, already sadly wilted—the "fresh plant material." Intended, no doubt, as an abstract—the current rage.

England: The exhibit to interpret a Shakespearean character, the name of the character to be printed on a card. When I couldn't get near enough to read the cards, the fun was to guess the characters. Often it wasn't hard. Juliet was represented by three red roses and a paper-cutter shaped like a dagger, and the balcony was an ivy vine with a mass of green podocarpus at the base. Shylock (or maybe it was Portia; I couldn't get near enough to see) was suggested by a postal scale (or maybe it was a fish scale), a heap of imitation gold-wrapped coins, and a branch of prickly Crown of Thorns, with its pink blossoms.

Peru—Lost Culture of the Incas: Required "a design using gold and other rich colors with dried, painted, or treated materials." The blue ribbon went to an entry involving gold-painted palm fronds, fungus, dried red celosia, and a figurine. A real beaut.

Italy—The Fountains of Rome: "Must give the illusion of water or use water as part of the composition, featuring fresh aquatic material." One entry consisted of a shallow oblong lead container holding half an inch of water, some green water plants, and the fleshless headbone of a catfish. The last was aquatic material, all right, but could hardly be called "fresh." Another lady, though, had ingeniously suggested a fountain by the curled, unfolding fronds of ferns, and the arrangement was pleasing. But is this arranging flowers? To me it's more like playing a game—a game little girls might play on a rainy afternoon.

France—Pomp and Elegance: Required roses and permitted accessories. Led to an absolute orgy of the latter—opera glasses, kid gloves, fans, lace, jewelry, yards and yards of artificial pearls. In contrast to the other classes, the bouquets were fairly opulent—usually a big vaseful of roses with gloves, pearls, fans, or whatever, ranged round the base. One was against a purple velvet background. I heard one spectator say as she passed a handsome jar of roses and other flowers, "That's the only thing she knows how to do—an absolutely *stuffed* vase."

Some exhibits were beautiful; the majority were modernistic, extremely studied designs. Old dried vines, I discovered, are particularly beloved for the abstracts. Design is now everything; flowers, as such, don't count. The old idea of bringing flowers into the house and preserving their natural beauty for several days in a bowl of water is forgotten. In the same show, the growing plants, not the arrangements, were what interested me; they were lovely, and they at least indicated that most of the exhibitors were capable of growing a good potted plant, a hanging basket, an orchid, or a cactus. The "Horticultural Classes"—i.e., specimen flowers of almost everything from a humble ageratum to a Eucharis lily or a Bonsai—showed that the ladies of the garden club also knew how to garden. How the exhibitors have time to grow anything, though, and still spend the days it must take to gather accessories is a mystery. Bad, cheap statuettes of animals, wrack from the beaches, pseudo-Oriental figurines, junk from the junk-jewelry counters, plunder from the drugstores, the notion counters, the wastelands—these were much in evidence in the exhibits. Some of the ladies must have travelled to distant lands for their exotic accessories—hemp and straw and grass mats from Tahiti, brass and copper pots from India, bits of carved ivory from Taiwan, a temple bell from Thailand.

The 1962 flower show in the same Florida city had as its theme "Horizons—Far & Near." Of the twenty-three classes under this heading, I particularly remember "Artistic Horizons," for which the rules read, "An interpretation of fresh and/or dried plant material, suggesting the painting of an artist. Staged before a thirty-inch circular background. Name of artist to be written on a small card." Of course the one representing Dali won, with a limp watch as its main feature. The best by far was Marcel Duchamp's "Nude Descending a Staircase," done in planes of cleverly clipped green material, such as palmetto and palm, and the stiff, swordlike leaves of the yucca. The woman who created this, a painter herself, knew her Duchamp painting and made a reasonable facsimile of it in plant material—a clever stunt and a handsome thing to look at, but again very far removed from the old idea of bringing the garden indoors.

In general, the flower shows of the South seem to go in for interpretations of more elaborate themes than do those of the North. Several years ago, a South Carolina friend sent me some pages clipped from show programs dating back to the late fifties and early sixties. I've saved them, because the themes of some of their classes delight me. One show had no general theme, but one of its classes in "artistic arrangement" had the general heading " 'Only in America,' Honoring Charlotte's Own Harry Golden," with these sub-classes: (a) The Statue of Liberty, (b) The Fourth of July, (c) Athens, Ga., 1891, The First Garden Club, (d) The Free Press, (e) The Declaration of Independence. Challenging, any way you look at it. I have pondered and pondered and haven't been able to decide what flowers I would choose to represent the Free Press. The Declaration of Independence would be easier: an arrangement of Bugle Weed or Achillea-the-Pearl or any of a number of other rebellious, takeover plants that submit poorly to control in a border would do very well. The ladies on the program committee were not thinking only of the American past, though. Another class in the show was titled " 'You've Never Seen Anything Like This Before!' Cape Canaveral—Outer Space."

The program of another show in the Carolinas did have a general theme, along with some pretty sketchy spelling and proofreading: "The Gardner's Twenty-third Psalm," which is apparently some gardener's paraphrase of David's beautiful poem. Each class in the show carried a quote from this pseudo-psalm. Class 1 was " 'The Lord is my Shepherd. . . .' An all white arrangement expressing an etheral quality"; Class 4 was " 'He giveth me wisdom when I become discouraged, only for the asking.' An interpretive arrangement featuring an accessory." Class 8 was " 'He sendeth the bees for a great purpose. . . .' An interpretive composition expressing God's plan for continual life."

A local annual iris show in the same region centered on dreams: " 'A Dream of Heaven'—composition depicting a spiritual quality; if all white, use a backdrop of colored velvet." "Day Dreams," "A Nightmare," " 'Pipe Dreams,' men's division"—you get the idea. I myself would have had most fun with that nightmare.

The flower shows of the North often have for their theme something as simple as "Dawn Till Dark," yet even a New England village show can follow a complex theme if the occasion really calls for it, as happened in 1958, when our neighboring town of Blue Hill was approaching the two-hundredth anniversary of its founding. The theme of the Blue Hill Garden Club's flower show that year was, naturally, "Blue Hill's Goodly Heritage," and the classes had sub-themes like "Pioneer Community Founders," with arrangements to be of the Early American type in pewter or tin containers, and "Benefactors," one of whom, Jonathan Fisher, the famous Congregational pastor, was remembered by eighteenth-century parlor bouquets. Under "Culture," two musicians who were early Blue Hill rusticators were honored

—Franz Kneisel, the distinguished violinist, by "A Symphony in Colors," and Ethelbert Nevin by arrangements featuring any of his compositions (his "Mighty Lak' a Rose" was a setup). One living writer, Mary Ellen Chase, a native of the town, was greeted by arrangements of "Flowers from the Bible," one of them a masterpiece of research. Blue Hill's history of industry was represented by four sub-classes: Shipbuilding (blue flowers in boat-shaped or oval containers), Lumbering (red flowers in wooden receptacles), Mining (nasturtiums in copper or brass), and Blueberries (berries and foliage in any container). I remember this show as one of the most delightful I have seen, running from period to modern arrangements, all of them simple and few of them farfetched or too literally interpretative. Kitchen utensils as containers are much favored in New England country shows, and any arrangement in an antique coffee mill is apt to take a ribbon. One did that year in Blue Hill. In small shows like this, the exhibitor at least appears to be having fun and to be less competitive in spirit than exhibitors I have observed at standard shows. Certainly he or she need follow less restricting directions and rules, and an air of relaxed pleasure prevails for participants and viewers alike.

New York City's annual International Flower Show is an affair quite different from the garden-club shows, which are all put on by amateurs. The International is open to florists, nurserymen, professionals of all kinds, as well as to amateurs, and it has numerous educational exhibits. Its exhibitors range from universities and botanical gardens and horticultural societies to the Federated Garden Clubs and the Girl Scouts. Flower arrangements are always there but do not predominate. The great beauty of the show is its gardens of seemingly growing plants, which are mounted by owners of the few remaining great estates that have professional gardeners, or by nurseries, botanical gardens, and commercial growers. (A great deal of floor space is given up to commerce and to sales of garden equipment.) When I lived in the city, I seldom failed to attend the shows, which were then held in the Grand Central Palace, in the dear, dead days when Bobbink & Atkins always seemed to carry off the grand prize for its garden display of roses. (I haven't ceased to mourn the demise of B. & A.'s famous rose catalogue.) In those days, too, the fun was to go early and see the lady amateurs installing their flower arrangements in niches. How simple and satisfying their pleasant vasefuls seem in retrospect, for naturalism prevailed, well before modern design and abstracts or thematic interpretation took over, and, of course, before show rules and regulations became so formidable. The last New York International I visited was the first to be held in the Coliseum, in 1957, and my chief memory is of its vastness, the fatigue it involved, and the difficulty of finding my way about. I did finally come upon, on the floor above the main exhibits, the sections allotted to *ikebana* demonstrations and to Occidental flower arrangements and made some irreverent notes on the latter. They had best be forgotten except for one scribble of summary: "Towers, triangles, shooting branches." My jaundiced summation of today's International Show's bouquets would surely be quite different, for even in the short time that has passed the styles have changed beyond recognition.

Books on flower arranging are a dime a dozen, or, to be more realistic, nearer to seventy-two dollars a dozen, at today's prices. I have been looking at them halfheartedly for years. A few I have kept because they are useful handbooks, and others because they are interesting in their graphic reflection of fads and trends. One of the few books on the subject that have endured at all (it is perhaps the only classic in its field) is

Gertrude Jekyll's *Flower Decoration in the House*, published in 1907 by Country Life and George Newnes, of London. The book, which lacks the range and literary quality of the more famous Jekyll books, is now out of print, but Scribners' 1964 Jekyll anthology, *On Gardening*, has a good section on "Flowers in the House," taken from *Flower Decoration* and some of the other less specialized books. Granting that *Flower Decoration* is written by an Englishwoman with a huge garden and a retinue of gardeners and household servants, the book still contains a lot of sensible advice for anyone—how and when to cut flowers in order to make them last, what sorts of containers are useful and attractive (both costly and cheap ones), and a discussion of foliage and flowers, season by season and month by month, with Miss Jekyll's ideas on which ones combine well. I don't agree with all her preferences in color, and although she was avant-garde for her period, some of her arrangements today seem rather heavily Victorian. I don't admire every vase of flowers, fruit, or foliage shown in the now rather old-fashioned photographic illustrations, but I do admire her approach, which avoids fussy props and adjuncts and is content with putting a few lovely flowers freely and easily into a vase that suits them. Many of her containers were collected in her travels; she was an artist as well as a gardener, and she designed a whole series of simple, inexpensive glass containers that were named "Munstead vases," after Munstead Wood, her home and garden. Most of her arrangements give me a sense of peace and harmony; they lack entirely the sense of effort and the self-consciousness so apparent in our contemporary flower-show exhibits.

If winning prizes at flower shows is your ambition, Gertrude Jekyll will be of little help today. Instead, turn to a book like *How to Make Cut Flowers Last*, by Victoria R. Kasperski (Barrows, 1956). The author was at that time the professional flower arranger at the historical monument of Mount Vernon and had made a study of all the factors that prolong the life of a bouquet. Hers is a helpful handbook. As things are going now, to win that garden-club Creativity Award an abstract arrangement may be your best bet, and you might turn to Emma Hodkinson Cyphers as your guide. She has written many books on modern arrangements and she is more open-minded about what Lafcadio Hearn would call the "vulgarities" of the past than some of the other writers on how to construct abstracts from garden trash. Her *Modern Abstract Flower Arrangements* (Hearthside, 1964) has chapter headings such as "Perspectives in Choosing Plant Materials *and Substitutes*" (the italics are mine) and "Symbols for Interpreting the Nonperceivable." The book is copiously illustrated with photographs of what to her, anyway, are abstract arrangements—some really good, some outlandish, and some, to me, not abstracts at all. My symbols and hers for perceiving what can't be seen may not be the same, but Mrs. Cyphers earnestly wants to keep you up to date, so I thought I'd better mention her. She can teach you how to make a stabile or a mobile out of something approximating plant material—balsa-wood discs sprinkled with caraway seeds, for instance, combined with dried pods and stems. *The Driftwood Book*, by Mary E. Thompson (Van Nostrand; enlarged second edition, 1966), is for those who like to combine driftwood with flowers. I don't happen to. If I must have driftwood in the house I prefer it by itself, if at all, as a piece of nature's sculpture, but this is a rather good book on how to collect, condition, and use driftwood if you can stand the wrack. If miniatures are your hobby, there's a really professional and thorough book by Lois Wilson, *Miniature Flower Arrangements and Plantings* (Van Nostrand, 1963). Mrs. Wilson's researches reach into antiquity and her illustrations in color and black-and-white of today's tiny bouquets are attractive. Her many appendixes giving lists of suitable plant material, names of nurseries where they may be obtained, and so on, will be helpful to the gardener as well as to the arranger. (As a child, I collected miniature china animals and miniature shoes from foreign countries. My meagre collections don't exist today, but I know now they would have been a treasure trove as accessories for an exhibitor of miniature bouquets.) If larger accessories are what you long for as an exhibitor, *Flower Design with Accessories*, by an Englishwoman named Merelle Soutar (Van Nostrand, 1964), is the best handbook I've run across. A few of the figurines shown in the illustrated arrangements seem to me dreadfully bad as art and thus ruinous to any flower display, but there are more good ones than bad ones, and they are all many cuts above those I saw in that Florida flower show.

I have never arranged flowers for a church, but the many who do will find good suggestions in a modern English book, *Arranging Church Flowers*, by Molly Purefoy, published in London by Pearson in 1964 and a year later in this country by Hearthside. The background of the book is very Church of England, but, barring a few special English Church rules, at least one of which—no flowers in the font—seems to have been made since the days of Francis Kilvert's "Easter Idyll," it is completely adaptable to any denomination. Its many plates are exemplary color reproductions and pleasing to look at, and its text might well be described by the book's subtitle, "Up to Heaven—Down to Earth." Almost every arrangement shown could be termed conventional, but each one is timeless and in excellent taste. For our recently built churches with modern architecture, something less floriferous and more dramatic might be needed. Mrs. Purefoy herself warns that the first thing to consider before drawing up a scheme for floral decorations is the architecture of your church. Her church was Tewkesbury Abbey, but she covers many others.

A non-book I found passingly amusing is *Flowers at the White House*, by a Washington newspaper columnist, Ruth Montgomery (Barrows, 1967). Its glimpses of the changing taste of our First Ladies over the years and its present-day tour of the restored famous rooms and their bouquets, now designed by the young genius Elmer M. Young, whom Mrs. Kennedy discovered and who is equally appreciated by Mrs. Johnson, provide the sort of gossipy information we all enjoy. There is always a cat-may-look-at-a-king fascination in looking at pictures of the smilax-festooned formal tables set for a state dinner and decorated with the bouquets of pink carnations that Mrs. Eisenhower liked so much, or the bowls of flowers arranged in the grand manner in antique porcelain and vermeil for the state visits of Haile Selassie and Jawaharlal Nehru, Konrad Adenauer, and Princess Margaret and Lord Snowdon when they were guests of the Kennedys or the Johnsons. The visiting dignitaries may have been too absorbed in weightier matters to notice them at all, but we can study them now not only out of curiosity but because they are modern, not consciously antiquated, arrangements suitable for their period settings.

Time and taste move onward, changing with the tides, and so does the aesthetics of flower arrangement. If one remembers that arranging flowers will always be a sub-art, not a true art, it doesn't really matter what style one experiments with. Flower arrangers are like yachtsmen; some prefer competition (yacht races and flower shows), some prefer sailing for its own sake (cruises and home bouquets). Eventually, as in fashions, skirts will come down again and elegance and simplicity, not oomph or splash, will take over in flower arrangement. My only hope is that freedom and flowers will be substituted for rules and flower substitutes and that everyone who likes to bring the garden indoors can again be "doing what comes naturally."

Finally, I shall confess to some of my own flower-arranging habits and rituals. I admit that, like the Japanese, I have *tokonomas*, but only in the sense that if I do not have wild flowers or something from my garden placed in certain spots in my home I feel as if the house were uninhabited, or, at best, sadly neglected. If I've happened to be successful with my bouquets, it is to those spots I look for a moment of refreshment or pleasure, but I cannot claim that I get spiritual inspiration from my hasty vasefuls, as the Japanese does from viewing his carefully constructed arrangement, placed in its single, ritual niche, or *tokonoma*. Like almost every housewife, I keep a bowl of flowers on the dining table, or at least an assortment of my own home-grown, hand-polished gourds or a flowering house plant. The silver tankard in the dining-room corner cupboard has its few roses in season, and the enclosed north porch, where we eat in summer, must have its small bowl of garden flowers or fall berries or whatever comes along. It is the living room I spend the most time on, for here there are always a pair of small arrangements in small pottery vases or antique, hand-painted, footed glasses of moderate size on the mantelpiece, flanking an oval family portrait. I always place a semi-miniature vase in a small niche in one bookcase—a foot-square shelf, too narrow for books, which makes a pleasing shallow recess that nicely frames a three- or four-inch vase of small flowers, like violas or violets, little wild tulips and daffodils, and grape hyacinths, lilies-of-the-valley, Star-of-Bethlehem, Allwood garden pinks, and so on throughout the summer. Unless a big flowering plant stands in one corner of a low bookcase next to the sofa, I put there a few luminous single hybrid peonies in a shallow dish in late spring, or, as summer comes on, I fill the biggest copper jar with decorative flowers—snapdragons, zinnias, sunflowers, huge African marigolds, larkspur, and chrysanthemums—or a tall old pale-green pickle jar that is just right to hold delphiniums, monkshood, hollyhocks, and the other very long-stemmed flowers. I use a minimum of pinholders and devices for supporting flowers, and have no chicken wire or florist's sponge for props. I follow no rules. If a vase can't support the flowers, I am apt to give up. I do use a simple pin- or glass-holder for flowers in shallow bowls or basins on the dining-room table, since there the bouquets must be seen from all four sides and must not be too tall. To sum it up, I am an indefatigable but careless and hasty arranger. It's a vice, all right, as Ogden Nash so perfectly expounded some years ago in his poem "The Solitude of Mr. Powers" and later in his book *You Can't Get There from Here* (Little, Brown). With his permission, I quote the first and last six lines of this devastating poem, which I hope has forever stayed my hand as it automatically creeps forward during a meal to right a fallen daffodil or nip off a wilting petunia:

Once there was a lonely man named Mr. Powers.
He was lonely because his wife fixed flowers.
Mr. Powers was a gallant husband, but whenever he wished to demonstrate his gallantry
His beloved was always out with six vases and a bunch of something or other in the pantry.
He got no conversation while they ate
Because she was always nipping dead blossoms off the centerpiece and piling them on her plate. . . .

Finally he said Hey!
I might as well be alone with myself as alone with a lot of vases that have to have their water replenished every day,

And he walked off into the dawn,
And his wife just kept on refilling vases and never
noticed that he was gone.
Beware of floral arrangements;
They lead to marital estrangements.

My own Mr. Powers is long-suffering, since he has a perpetual case of pollinosis as well as a flower arranger to contend with. If he walks off into the dawn some fine morning, I shall know what and whom to blame.

12

Winter Reading, Winter Dreams

New Marguerite carnations. From "Harris' Rural Annual," 1897

December 16, 1967

The year 1967 started with an all-out alert on the danger of poisonous plants. On January 6th, the *Times* published a story about a lecture on the subject by John M. Kingsbury, who is the author of a useful small book titled *Deadly Harvest: A Guide to Common Poisonous Plants*, illustrated by identifying line drawings and photographs (Holt, Rinehart & Winston, 1965). The *Times* reporter drew heavily on Dr. Kingsbury's compact *Guide* for her rather naïvely astonished recital of dangers to children in their mothers' flower beds. Later in January, Delacorte published a picture book with drawings by Leonard Baskin and text by Esther Baskin, titled *The Poppy and Other Deadly Plants*. The two books could hardly be more different. The first is practical and scientific; the second is romantic, literary, a bit vague as to facts, but a work of art. The *Times* story was no news to me or to any gardener, botanist, farmer, or sheep or cattle grower, each one of whom has long known that many of our most beautiful flowers, trees, shrubs, and weeds, both wild and cultivated, are poisonous if a human being or an animal ingests the roots, the leaves, the flowers, the berries or fruits —whatever part the poison happens to lie in. We all know, too, that many other plants are not killers but are poisonous to touch (poison ivy, poison sumac, poison oak, and so on), and that still others are allergens for certain individuals. Dr. Kingsbury covers all types of poisons quite thoroughly. A professor at the State College of Agriculture at Cornell, he is a specialist on the subject, and his technical book for scientists is *Poisonous Plants of the United States and Canada* (Prentice-Hall, 1964). His less expensive 1965 handbook for popular consumption is an excellent guide, especially for parents and for camp directors, many of whom now seem to like to send their children on "survival trips," in which the campers live off the land and water they cover. Dr. Kingsbury's *Guide*, if followed with common sense and no panic, can prevent needless deaths of children, heedless adults, and many farm creatures. He finally says, or implies—and I agree with him—that there is no real protection for very young children except supervision and education and eradication of special dangers near or in their homes. They should be taught never to eat leaves and roots and flowers or to swallow a berry without an adult's permission, and as soon as possible they should be educated to recognize what plants are to be avoided completely. At a very early age, I remember, I was taught how to recognize and stay away from deadly nightshade, poison ivy, and poison sumac. (I was just as early taught the delights of chewing tender young checkerberry leaves and sassafras root.) To me it would be ridiculous, though, not to grow monkshood, foxglove, hellebore, larkspur, autumn crocus, poppies, lilies of the valley, buttercups, and many other flowers now present in my borders just because they have some poison in them (in juice, root, bulb, seed, or flower) that can be injurious under certain conditions—most often, of course, if it is swallowed. Even young potato sprouts—and potatoes themselves, if their skin is green—are highly poisonous, while just outside the home grounds grow still more dangerous plants, like the vicious water hemlock and the deadly nightshade, in roadside ditches, meadows, swamps, woods, or suburban empty lots. Garden shrubs or plants with poisonous decorative berries are a different matter. The bright berries of some varieties of Daphne and yew, the shiny bean of the castor bean—no sensible mother will grow these plants, since very little children instinctively reach for a gleaming tidbit to pop into their mouths. One safeguard, of course, is to cut off poisonous fruits, seedpods, and berries the moment they appear, but this isn't a sure precaution. It *was* news to me that the hardly noticeable flower of the poinsettia is highly poisonous if it is eaten. Its surrounding Christmas-red so-called petals (actually bracts), which might attract a baby, are also poisonous, so if you have the kind of child who is a nibbler of bright objects or who would pluck a huge poinsettia bloom and chew it up—bracts, blossoms, and all—beware! In the South, picnickers must learn never to grill meat on a twig of oleander, and so on. Knowledge is the answer, and Dr. Kingsbury gives many warnings for odd situations. The Baskins' handsome volume of deadly plants is a graphic delight. The text is full of interesting myths, literary references, and facts from ancient herbals. It covers only a few of the killer poisonous plants, and I would never count on it as a guide, but I enjoyed it.

I feel sympathetic to country children who like to forage for edibles, because I liked to myself, and my youngest grandson, now eight years old, still thrives on a diet of green fruit and berries. He seems to prefer a plum or apple he has picked too soon to windfalls or ripened fruits that have been harvested. A sense of guilt about this oppresses me, because it could well be that he inherited his scavenger habits from me; I am thinking of the long autumn walks my next-older sister and I took, when I was his age, in the lanes, orchards, and spinneys of Woodstock, Connecticut, and when the game was to see how many nuts, apples, peaches, berries, and unharvested tomatoes and carrots left in the vegetable rows we could stow away before we returned to the tame harbor of our aunt's garden and orchard.

When I was even younger, I had a solitary adventure in the same locale that I might confess to here, ludicrous as it was, since it involved both Woodstock and a Victorian garden-centered children's book. I was visiting my aunt, this time alone, and to keep me occupied she gave me an even then old-fashioned children's book—Juliana Horatia Ewing's *Mary's*

Meadow, which was published in London by the Society for Promoting Christian Knowledge in 1886. I have reread it recently and wonder now why I found it so absorbing, since it is sentimental, moralizing, and far removed from an American child's habits of mind and life. It is even hard to believe that English children with a family gardener were as interested in topsoil, compost, and very special flowers as Mrs. Ewing's young family. However, I had read and liked *Jackanapes*, Mrs. Ewing's juvenile classic, and I snapped up *Mary's Meadow*. The story was about a large, lonely, bookish family of English children whose passion was for gardening and flowers. Mary, the narrator and the oldest child—in her mother's absence the Little Mother of the brood—dug out of the family library John Parkinson's famous seventeenth-century *Paradisi in Sole Paradisus Terrestris*, and to amuse her brothers and sisters she invented a game in which the children created an "Earthly Paradise" of their own. Each character had a name derived from books, flowers, or gardening and a part to play in their game. Mary's was Traveller's Joy, who sows seeds and plants flowers from her home garden in the hedgerows and roadsides to give joy to passersby—a sort of British Johnny Appleseed. Her title in the game was derived from the British common name for the wild clematis that brightens the hedgerows of the English countryside, but I had never heard of either Traveller's Joy or clematis at that age and I read the title symbolically, as did the children of the book in their game. Thus, one day, when my aunt was dividing her garden perennials, I gathered up a few of her discarded roots and started out with a trowel, determined to be Woodstock's Traveller's Joy. The borders of the Connecticut dirt roads, where then the chief traffic was horse-drawn, seemed too full of bloom and too public and too difficult to dig in for my surreptitious enterprise, so I turned up a grassy cow lane belonging to a neighboring farmer, and there I hastily planted my garden flowers. I even went back to water them the next day. It did not occur to me until years later that the lane must have been alive with the bloom of far rarer and more lovely wild flowers than my garden transplants of common garden phlox and Golden Glow or that there would be few travellers up the lane except Mr. Sumner, the farmer, his cows, and his daughters, who were my friends. But all the next winter I took a secret pleasure in my comical plantings, which, so far as I know, did not survive. Great is the power of books!

The first significant American book on fragrance in the garden since 1932, when Louise Beebe Wilder's memorable work, *The Fragrant Path*, was published, is *The Fragrant Year*, by Helen Van Pelt Wilson and Léonie Bell (Barrows, 1967). It is a beautiful book, illustrated with a hundred of Mrs. Bell's exquisite pencil drawings, which to me are the best part of this handsome volume. The text is workmanlike and the authors are sincere in their appreciations of the multitude of garden scents, but in spite of several fine passages I found the text a bit disappointing. It sent me back to reading Mrs. Wilder, who has always charmed me, and to another famous study of fragrance—*The Scented Garden*, by Eleanour Sinclair Rohde, an English book first published here in 1931. Both these writers are now rather out of date on horticultural varieties and nomenclature, but their books can be read right through with pleasure as books rather than as reference books. The authors of *The Fragrant Year* have produced an excellent and even imaginative reference book and have covered the subject of fragrance plant by plant and month by month, including house plants, shrubs, bulbs, a few trees, and many flowers. They draw on their own experience in their two home gardens in Westport, Connecticut, and Conshohocken, Pennsylvania, which happen to have similar climatic conditions. The book is written in the first person plural—a pronoun that in this case is unsatisfactory and occasionally confusing. Mrs. Wilson's introduction refers to the "we" as the editorial "we," but it is not that at all. It is the plural "we." I couldn't always be sure where I was geographically or which expert was speaking. I wish the authors had divided up the chapters and written them separately, referring back and forth to each other's garden when need be and indicating which author wrote which chapters. Much has been made of their elaborate reclassification of scents—an undertaking I find highly dubious, since the nose is such an unreliable organ and varies so much from individual to individual. A few of their nine main categories of fragrance make sense to any nose, such as *Balsamic*, *Heavy*, *Spicy*, and *Fruited*, which has four subdivisions. But what about the categories *Sweet* and *Honeyed*? The latter is divided into (a) "dry, musty, almondlike, sweet," (b) "yeasty, from sweet to offensively strong," (c) "musky, fermented, sweet." The catchall word "sweet" could apply to most scents, and the distinction between *Sweet*, which is one of their categories, and "sweet," the adjective describing *Honeyed*, is a bit odd. What, too, about the category *Unique*, under which they list only lily of the valley, sweet pea, some irises, wisteria, and common lilac? To my nose, admittedly now unreliable because of years of smoking, heliotrope is heliotrope and unique, and so are mignonette and *Clematis paniculata*, carnations, and countless other flowers. The fragrance of violets, they later admit in a separate chapter, *is* unique, and they make *Violet* a fragrance category, as they do *Rose*. The chapter on roses, obviously a labor of love, is one of the most rewarding—on fragrance, on species and varieties, and on the history of the rose strains.

There are some notable omissions, especially among

the tulips, which are, as they say, mostly unscented, but there are quite a few more with fragrance than they list. (The delicious parrot tulip Orange Favourite is one, and there are Ellen Wilmott and others among the modern hybrid tulips.) The authors also, I feel, give too few warnings against the seeds of new hybrid flowers that have lost the scent of the original strain, and too few warnings against oppressively fragrant plants. However, the nose does differ on the latter matter; what sickens one person will delight another, as I have found in my own household with some kinds of lilies and with the—to me—deliciously scented waxplant. As a reference book, *The Fragrant Year* also lacks that essential ingredient for the practicing gardener—the names and addresses of nurseries and seed houses that are sources for some of the rarer scented plants. Despite these qualifications, get the book. No gardener who cares for fragrance should be without it, and no public library. As for the illustrations, my admiration is unlimited; Mrs. Bell is an outstanding botanical artist, with the ability to make her pale-black pencil drawings create texture and character, and sometimes—magically, or so it seems—even color. Her detail is amazing, yet her drawings are works of art; for accuracy, they surpass any color photograph and most of the black-and-white textbook botanical drawings I have seen of late, in an era when botanical art is at low ebb.

This year, it is the wild-flower enthusiasts of the Southeastern states who are in luck, because of the publication of Volume II (in two huge books) of the New York Botanical Garden's *Wild Flowers of the United States*, an unprecedented and mammoth work-in-progress, of which Harold W. Rickett, the Garden's senior curator of botany, is the editor and McGraw-Hill the publisher. It costs more than last year's two big books, on the wild flowers of the Northeast, and understandably, since Volume II (six hundred and eighty-eight pages in all) is much longer than the two parts of Volume I. It lists nineteen hundred species and is illustrated with sixteen hundred and ninety-five full-color photographs and three hundred and eighty-four line drawings. The botanical introduction and guide to the families of plants of Volume I is repeated, with its charts and explanatory line drawings. The two new books include not only more text but many, many more flowers, which require many more new identifying color plates. The states covered this time are the Carolinas, Tennessee, Georgia, Alabama, Mississippi, Arkansas, Florida, Louisiana, and the eastern and coastal sections of Texas—a wide range so far as climate and soil go, embracing woodlands, meadows, swamps, and parts of three mountain ranges (the Appalachians, the Blue Ridge, and the Ozarks), not to speak of the unique Everglades. All the flowers that grow in both the Northeast and the Southeast are shown again (and there *is* considerable overlapping), occasionally with a better photograph than the one in the first volume or with minor corrections in text or botanical classification. I noticed, too, that some of the color plates repeated from Volume I appear to be improved in color *reproduction.* These ambitious books, which describe far more wildings than have ever before been put between covers in a book for the average reader, grow bigger and better as the series progresses. Dr. Rickett has more collaborators this time, and he states in his preface that it is "with trepidation" that he presents the flora of "a vast and varied region" with which he is "only slightly acquainted." With how much more trepidation, then, do I, an amateur who is only moderately well acquainted with the wild flowers of New England, New York, and the Main Line region of Pennsylvania, attempt to review these important books! I can do it only in a general way, not as an informed review. I have spent a good many winters on the west coast of Florida, but I have never found the region inviting for walks in the wild, except along the beaches. Poisonous snakes lie in wait, knife-sharp palmetto and prickly scrub tear at your legs, and bayous and islands and sandy keys are bordered with impenetrable if glorious mangrove barricades. However, in the state and national parks and preserves in which Florida abounds, a Northerner comes upon unknown and delicious wild flowers, and the sides of the roads off the main highways suddenly burst into brilliant weedy flowers he does not at once recognize. The new books go South with me this winter, and I hope they help me widen my acquaintance with Florida wild flowers. The "Orchid Family" section of the new books is especially beautiful in its photographs, and it especially fascinates a New Englander used to the few noble but sparse orchids of a cold climate. Dr. Rickett acknowledges a debt to Dr. Carlyle Luer, of Sarasota (which happens to be my Florida city), for both the text and the photographs of the "Orchid Family."

I must reluctantly add that the built-in drawbacks of Volume I that bothered me a year ago still persist in Volume II—photographs instead of more exact and lifelike botanical paintings; books too heavy to hold in the hand, far worse in this respect, of course, this year; again, an index only at the end of Part II instead of including the index at the end of each big book. This means that for viewing, reading, or consultation for identification one must own a double-width dictionary lectern, sit at an oversize table, or else resort, as I do, to a wide double bed with lots of propping pillows for the big tomes, since both books have to be available, side by side, for consultation of the index. Despite the cumbersome format, I can testify that in the past year I, an average reader unversed in botany, have found Volume I definitely useful and readable. Volume I's two sections are bound in a gay green buckram; the two for the Southeast are clad in muddy oxford gray. Obviously, the editors do not expect most readers to buy all the volumes of this vast undertaking as a matching set. Last year, I expressed the hope that this great scientific work would eventually be broken down into smaller, handier, less expensive state-by-state reference books, and I am happy to report that Dr. Rickett has written me that this is the eventual plan, but he says that its execution will be in the far future. The important thing now is to have the rest of the big volumes adequately financed and the entire United States covered, so that its enormous wealth of wild flowers may be revealed to every interested citizen.

A new book to browse about in for the sheer entertainment of its astonishing facts and its copious photographic illustrations is Edwin A. Menninger's *Fantastic Trees* (Viking). The author, who is an authority on trees and tropical horticulture, addresses his prologue to his psychiatrist brother Dr. William C. Menninger, of the Menninger Foundation in Topeka, pointing out that trees often seem to behave like human beings and have many of the ailments, aberrations, and eccentricities of men. The book is really a series of case histories of the "marvels and monstrosities of the arboreal world." Obesity and gigantism in trees, trees that switch sex, trees that are gregarious and trees that must grow in solitude, even self-sacrificing trees are just a few of the curiosities he cites. Most interesting of all to me is the section on "Trees That Cannot Live Without Animals." There is, for instance, a tree whose blossoms are pollinated only by bats, and another pollinated only by rats, and there are tree seeds that will not germinate unless they have been through the digestive system of a giant tortoise. Other seeds have to be disseminated and fertilized by elephants if they are to sprout. These are but a few of the oddities of reproduction and survival. One chapter, titled "The Wasp Trees," is so complex and surprising that it could provide conversation enough to revive a lagging six-course dinner party.

Some Ancient Gentlemen, by Tyler Whittle (Taplinger, 1966), is one of those rare books with garden subject matter that can be read right through for the pure pleasure of their literary style, their civilized wit, and their extensive knowledge of global garden history and literature. Mr. Whittle is also a do-it-yourself gardener and a botanical collector. He amusingly recounts the pleasures and pains of his own unorthodox English garden and the hazards of handymen helpers. What makes me happiest, though, is his consistent humor and his tongue-in-cheek comments on the fidgety habits and sheer lunacy of the dedicated gardener. Many should find his wide-ranging discourse a lifesaver in the midst of the pedestrian or overdecorated garden books that American publishers too frequently favor. The book is illustrated modestly with decorative line drawings, photographs, and a few reproductions of ancient garden engravings and plans.

The Rothschild Rhododendrons: A Record of the Gardens at Exbury, by C. E. Lucas Phillips and Peter N. Barber (Dodd, Mead), will make good reading and viewing for all growers of rhododendrons and azaleas. Their name is legion today, both North and South, now that suburbanites and average gardeners as well as the owners of great estates are able to grow these formerly exigent decorative shrubs. The big volume is also practical, since it contains the 1966 "Exbury Register" of rhododendrons and azaleas developed by Lionel de Rothschild and his son Edmund at Exbury, many of which are now among the most prized varieties because of their bred-in hardiness and their unusual beauty. There is also an excellent chapter on their culture. The "Register" gives their ancestry and their British and American ratings, with notations on the varieties that cannot stand cold weather. All this sounds like a book for the connoisseur only, but for an amateur gardener like me, who does not grow rhododendrons or azaleas, it makes good reading, too. Its narrative chapters tell the unusual life story of Lionel de Rothschild, who was not only the head of the great Rothschild-family London banking firm but also one of the most gifted horticulturists of this century, and they describe the gradual development from near-to-scratch of Exbury House and its surrounding two hundred and fifty acres of gardens and woods on the Beaulieu River, near Southampton. "Mr. Lionel," as he was known to his staff, would spend the week in London, dash down to Hampshire in his Rolls-Royce two-seater every Friday, and there direct each detail of the huge enterprise. He himself marked the spot where every tree, shrub, and plant was to be placed, and he himself did most of the hand-pollination that achieved the famous hybrids. The First World War delayed his work, but between the two World Wars he grew the finest rhododendron garden in the world—an estimated million plants—and in the twenty years he had at Exbury before his death he made twelve hundred and ten crosses, from which four hundred and sixty-two new varieties were named and registered—an amazing achievement, horticulturally, in so brief a time.

The story told here is the record, too, of one of the last examples of English gardening in the grand manner. At the start, there was a task force of a hundred and fifty men to dig two feet deep into every inch of the ground in the Exbury Home Wood, where the rhododendrons were to grow. Ponds were drained and their bottoms cemented, and new ponds were made. There were sixty gardeners to work in the woods and fifteen in the greenhouses, a clerk of the works, and, presumably, other clerks to keep the records and compile the all-important studbook. Yes, it is called exactly that; a prize rhododendron's heritage chart is even more complex than that of the bloodlines of a race horse. The color plates, from photographs by Harry Smith, comprise half the large volume, and despite a few minor failures in green and blue tones they are superb. Mr. Phillips writes with warmth about Lionel de Rothschild and with clarity on horticulture; his collaborator on the text, Peter Barber, is the present director of Exbury Gardens, now a commercial nursery that continues to develop more hybrids. The estate has been reduced in part to farmland and in part to the public nursery, which is owned by Edmund de Rothschild, but the great rhododendron wood is still preserved and is private—not a part of the National Trust. Edmund de Rothschild no longer lives in Exbury House but at nearby Inchmery, the home of his infancy, where his father lived while he was laying out Exbury. The story of the diminishment of the great Exbury estate by the Second World War and a changing economy is told sadly and forthrightly in the book. Unfortunately for Americans, the facts of how best to obtain named varieties of Exbury-strain rhododendrons and azaleas in this country are not explained, since the book was first published in England, for an English audience. Exbury varieties or strains are available at a number of our good general or specialized nurseries, and are listed under slightly varying names (involving the words "Exbury" and "Rothschild") and often with conflicting and sometimes ambiguous guarantees. Some of the general nurseries list unnamed varieties and only in mixed colors. If I were buying, I would order from a specialized nursery that is explicit on variety names and on hardiness, since no gardener should be so foolish as to buy mixed colors or plants not suitable to his climate zone. If he does, disappointment lies ahead. I remember only too vividly my father's chagrin after he finally invested a considerable sum in a small plantation of rhododendrons in colors he had been at some pains to select. The bushes all lived and appeared to thrive, but every one of them produced only pallid lavender-white flower trusses each year, and every June my father suffered and muttered. He had been gypped and he knew it, but at that he might have been worse off if his faithless nurseryman had sold him indiscriminately mixed colors and he had wound up with a clash of purple and orange blooms against the pink brick wall of our house in Brookline. This beautiful volume would have given him many happy winter dreams, as it may give them to others today, just as it has to me. After all, winter reading and winter daydreams of what might be—the gardens of the mind—are as rewarding a part of gardening as the partial successes of a good summer of bloom.

13

Winterthur and Winter Book Fare

Magnolia. From Breck's, 1884

December 21, 1968

Guidebooks and field guides sound like dull fare, but this year it happens that some of the best books for those interested in flowers, whether wild or cultivated, come under this heading. One of them is among the most beautiful and readable of all the 1968 garden books. It is *The Gardens of Winterthur in All Seasons*, by Harold Bruce, with photographic illustrations in color and in black-and-white by Gottlieb and Hilda Hampfler (Viking). The printing and all the plates, both the black-and-whites and the exceptionally fine color ones, were done in Switzerland, and with near-perfect results. Unless you are lucky enough, though, to turn up an unsold copy in a bookstore, you will not be able to get the book at once, for it turns out that the Viking first edition, of a meagre thirty-five hundred copies, has already sold out. There is always, however, the handsome paperback edition, which I describe later, and one can hope that a second printing of the more durable Viking book will soon be available.

As almost every reader must now know, Winterthur is the name of the great du Pont estate three miles from Wilmington, Delaware. Over a century old, Winterthur is now an estate of nearly a thousand acres, of which sixty comprise the present gardens. They are open to the public from April through June and for about six weeks in the fall, and are visited each year by many thousands. When one considers the photographs of the gardens in winter and summer, it seems too bad that one cannot wander along these lovely paths all year long. The museum, once the family mansion and still within the gardens' boundaries, houses the du Pont collection of decorative objects and furniture made in or imported to America during the seventeenth, eighteenth, and nineteenth centuries; it began as a family collection of fine antique American household furnishings, acquired generation by generation, but was enormously enlarged by the present owner, Henry Francis du Pont. He was the first du Pont to open the gardens to the public, in 1951. The museum may be visited all year long, but arrangements for a ticket of admission must be made beforehand.

The book starts with a brief foreword by Mr. du Pont and a short "History of the Gardens" by C. Gordon Tyrrell, their present director. From this history one learns that Winterthur came into existence in 1839, when James Antoine Bidermann and his wife, Evelina Gabrielle, the daughter of E. I. du Pont de Nemours (born in France but settled near Wilmington in 1802), acquired four hundred and fifty acres of land near the estate of Mrs. Bidermann's father. After clearing the woods, the young couple built *their* own home in the New World and named it Winterthur, for the town in Switzerland where Mr. Bidermann's family had lived. The sunken garden Mr. and Mrs. Bidermann started near their home no longer exists, but the gardens were enlarged in 1875 by Mrs. Bidermann's nephew, Colonel Henry A. du Pont, the then owner, and from then on, under the Colonel and his son, came the miracles of landscaping and planting and planning that only good taste and vast money could have created. In effect, the du Ponts have given to the United States, for at least part of the year, a huge privately owned park, with an arboretum, research buildings, greenhouses, a pavilion, and a museum. If you can't get to Winterthur, Harold Bruce's book and the Hampfler photographs will make you feel you have been there.

The book is also published in a paperback edition as *Winterthur in Bloom*. The one thing I prefer about this slightly cheaper edition, which is identical with the Viking one except for the binding and cover, is the photograph chosen to decorate its glossy, hard-paper cover, which shows in brilliant color the very early *Iris reticulata* Royal Blue, its green stems and blue blossoms rising, undaunted, through a light spring fall of snow. I like paperbacks, but this book is too heavy and too enduring in value to be put into paperback; already the spine of my copy is torn and weak, and, like human spines, it will crumble with use.

The Viking edition is jacketed by another remarkable Hampfler color photograph, showing in the background a narrow vista of pond, road, a harvested field, fields of very green grass, and, in the far distance, green hills and trees. In the foreground, and dominating the scene, are two ancient russet-brown trees—a beech and a hickory, their leaves just starting to fall. To my New England eyes, used to a blaze of gold and red in late October before the killing frosts, the picture seems a bit melancholy. It reminds me of our very-late-fall russet-and-evergreen Maine landscape, when the deciduous hackmatacks are losing their yellowing needles, the brilliant maples and the elms are bare of leaves, and the long-lasting mulberry-red foliage of the pear tree and the mulberry-pink patches of low-bush blueberry on the hills of late October have turned to a dismal brown. To a Maine resident like myself, this means that the beautiful long cold white winter I used to look forward to, in spite of its problems and hardships, is about to set in, bringing suffering to our sick or elderly neighbors. This is, of course, purely a personal, emotional reaction; gold and brown are right for autumn in Delaware and the Zone 6 climate of Winterthur. The temperatures—cold in winter, hot in summer, and midway between northern and southern conditions—are one reason for the extraordinary variety of trees, shrubs, and flowers that can be grown there.

All in all, the photographs of the Hampflers are, as Patrick Synge writes in the *Journal of the Royal Horticultural Society*, "absolutely superb and have given us one of the finest illustrated garden books that I have seen for a long time." I agree entirely. (There is, by the way, no way of knowing which of these remarkable pictures were taken by Gottlieb and which by Hilda.) Mr. Hampfler and his wife spent a year at Winterthur making five hundred prints, from which a hundred and sixteen black-and-white and thirty-two color photographs were selected. My favorite page of all—not in color—is of a clump of single white Japanese iris just coming into bloom; its slim, elegantly pointed buds and strong stems are of varying height, and its two first flowers rise above narrow, rush-like leaves.

For lack of garden space, I have never grown Japanese iris (*Iris kaempferi*)—and, too, because the nurseries now seem to offer only the double variety, which to me seem less like iris each year, as they get further and further away from the purity of form and exquisite delicacy of the singles. If there is a nursery that has tucked away but does not list roots of the single whites, I hope it will speak up, for at last I have a place to plant two big clumps. A year ago, we damned a small brook on our farm, and we now have a small, rather deep pond at the edge of a wood where alders and wild apples overhang. We want to keep the pond simple and plant very little—a native cattail or two, perhaps, and a few wild flag from our swamp—and leave the rest to nature, and, we hope, to trout. However, I had had my heart set on those two stands of single white iris (planted not *in* but *near* the water, on the south side of the pond), if only because they would be visible above the vegetable garden and north hayfield from our house and terrace. Our only planting around the pond so far has been of crown vetch, to hold the steep banks at the sides of the cement dam. Next spring, I may try to sneak into the grass on the steeper north bank of the pond some of the violets that spread like weeds under my shrub roses. They might help retain the bank if ever again we get a thoroughly wet summer.

To return to the book, the Bruce text, like the photographs, is superior. Mr. Bruce was the taxonomist at Winterthur when he started his book—i.e., a specialist in plant classification and nomenclature—but his prose is far removed from the turgid writing of most horticultural scientists. He has divided his story into four sections—Winter, Spring, Summer, Autumn—and he leads one through the seasons at Winterthur in a flexible, non-botanical, easy prose style, obviously written *con amore* and with a deep appreciation of what the natural world and the world of the landscaper and gardener are all about. He hasn't attempted to write a botanical handbook, and, as a taxonomist, he has had the courage to change the rules of the gobbledygook of current botanical Latin, when he must use it for identification, by capitalizing the lower-case botanical adjectives based on proper nouns—such as *smithii*, *sargentii*, *tomasinianus*. Where it seems called for, he tells us who Smith and Sargent and the others were. He also often translates the Latin names of species and explains their origin, and thus his story of Winterthur is full of history, mythology, and legend. Common English names get the same attention on origin, and he is careful to point out the regional character of these names; "Spring Beauty" and "bluebells" mean quite different plants in different parts of the country. Along the way he gives expert advice on mass planting, and practical advice for the less ambitious gardener on the horticultural requirements of each tree, bush, plant, bulb, and corm he describes. I have learned a lot about the small home garden from Harold Bruce, and I am especially grateful for his notes on hardiness, for I have been nursing along plants that will never flourish in our own Zone 4 to 5.

My fear for the Winterthur of today is that zeal, experiment, and expansion may lead to overplanting. Enough is enough, and a few of the photographs seem to me to point to too much, especially the one of the golf course on the estate. If I were a golfer, I would be distracted by all those naturalized daffodils following the elegant curves and making patterns of irregular spheroids all over the edges of the fairway; how could a pull or a slice be retrieved without ruining the flowers? And what was once a quarry seems in its photograph to show little sign of a natural quarry, because so many new rocks have been added, to turn the quarry into a floriferous rock garden. It is now completely filled with plants, and it has ornamental iron fences and resting outlooks to make it safe and pleasant for throngs of visitors. The author says, "None of these views is accidental." How sad, I thought—not a natural vista left, not a natural quarry, with its cattails and spooky walls, dripping water, and its deep pond. But I am only an armchair visitor, and if I had followed the many walks (there is a map) I might well feel differently.

Among the great charms of the du Pont estate are its age, its ancient trees, and—in many of the earlier plantings in the Azalea Woods and elsewhere—the now nearly lost beautiful early varieties of azaleas and cherries, lilacs, crabapples, and so on, obtained by Colonel du Pont and his son through their close friendship, in the early days of Harvard's Arnold Arboretum, with its trio of great plantsmen: Charles S. Sargent, E. H. (Chinese) Wilson, and Karl Sax. Later, in the thirties, quantity buying for Winterthur was done through private sources, and whole nurseries and private gardens were purchased. Many of the varieties so acquired are now also unobtainable. This means that the book cannot offer that helpful ingredient in a volume of this sort—a list of nurseries and other sources of supply. It explains, of course, why I cannot get my single white Japanese iris roots. Well, one can't have everything.

A welcome volume in the Peterson Field Guide series is *A Field Guide to Wildflowers of Northeastern and North-Central North America*, with text by Roger Tory Peterson and Margaret McKenny and illustrations by Mr. Peterson in color and line drawing (Houghton Mifflin). It is the best try yet, perhaps, at a pocket, fairly comprehensive, compact volume covering a large region, but it is less compact and comprehensive than Peterson's famous *Field Guide to the Birds*, which covered the United States east of the Rockies. This was inevitable, since wild flowers are more complex and numerous than birds. The subtitle of the new volume is "A Visual Approach, Arranged by Color, Form, and Detail." Like Mrs. Dana, of our childhood, in *How to Know the Wild Flowers*, Mr. Peterson, to help the amateur, groups his flowers by color rather than by family, and attempts to provide for instant visual recognition by his illustrations. Unhappily, only a small percentage of the illustrations are in color, and not very good color at that. Unlike the flowers themselves, this book does not charm the eye or, like Mrs. Dana's book, the mind. The very first flower I tried to look up in it—a big, dandelion-like Goatsbeard, sent me from upper New York State—is missing, but I did find it under *Tragopogon major*, with exact description and two illustrations, in Harold W. Rickett's huge *Wild Flowers of the Northeastern States*, sponsored by the New York Botanical Garden. Such omissions are inescapable in compact guides. Mr. Peterson, who had agreed merely to edit this book and write the legends for the illustrations, ended up by having to write half of the text and do all the illustration himself—a major labor of love. His hundreds of line drawings are excellent. His introduction and particularly his page called "Survival" were especially interesting to me. To save space, the flower families are designated by symbols, but there are thirteen pages of these tiny complex symbols. I, for one, could never learn to recognize them at a glance; some are too fancy and others are too similar. Most amateurs, though, don't care very much about flower families, and almost thirteen hundred species, covering both the Northeast and the North Central states, is enough for a pocket field guide. The book should be in every household or library, but because it is fat it will stretch many a field-jacket pocket.

The University of Tennessee Press last year brought out two really handsome hard-paper-cover *little* spiral-bound books—an enlarged edition of its earlier *Great Smoky Mountains Wildflowers* and a new edition of *Rhode Island Wildflowers*. The illustrations in both books are delightfully good color photographs; the flowers covered are very limited in number, but what there are are so alive that they seem to be growing on the page. This is the first time I have found myself actually *liking* color-photograph illustrations of flowers —aside from, of course, the Winterthur book, and a few other de-luxe books, such as some of the Sierra Club series.

Another new regional guide for wild flowers is *Wildflowers of Cape Cod*, a paperback by Harold R. Hinds and Wilfred A. Hathaway, illustrated by clear line drawings by Hathaway and a few none-too-good photographs in color by the authors. (The book is published by the Chatham Press, in cooperation with the Cape Cod National Seashore Park.) The guide has an interesting and helpful scheme: the flowers are grouped not by color or by plant families but by eight habitats—woodlands, dunes, boglands, fresh-water meadows, etc. Six hundred species are covered. Warnings are given about the most poisonous wild flowers. For Cape Codders, this should be a useful if not handsome local handbook.

The third volume of the New York Botanical Garden's vast work-in-progress, *Wild Flowers of the United States*, is not to be issued until the fall of 1969. Each volume comes in two huge books, as you will remember, and Volume III, says its publisher (McGraw-Hill), will be on the state of Texas alone, and will be less weighty and easier to read. The editor, Harold Rickett, is the senior curator of botany at the New York Botanical Garden. Listing every known wild flower in the fifty states is bound to bring delays. However, it is the only guide that really deserves the adjective "comprehensive," at least in the two regions covered thus far—the Northeastern and Southeastern states.

Rock Gardening: A Guide to Growing Alpines and Other Wild Flowers in the American Garden, by H. Lincoln Foster, with line drawings by Laura Louise Foster (Houghton Mifflin), is not really a guide at all; it is an authoritative, four-hundred-odd-page reference book on this subject. Mr. Foster explains the uses of and methods for establishing rock gardens of all sorts in our climate, but the body of the book is an alphabetical list of suitable plants, with directions for growing each of them. The book makes a useful American adjunct to the classic work in this field—Reginald Farrer's two-volume (over a thousand pages) *The English Rock-Garden*, which, unobtainable in this country, is still available in England, where it is a standard reference book, even though it was first published in 1919. It is delightful to read anywhere. Farrer was many things—a scholar, a versatile and exciting writer, an eccentric, and, above all, a plant explorer, who collected all over the world, mostly in snow and ice. He was also a practical gardener and a designer of rock gardens during his later years in England. His explorations and collecting were mostly done at the end of the nineteenth century, but he has forever left his mark on gardening by his plant discoveries and travels and by his rock-garden book.

My own attempts at rock gardening are minimal: two small patches of plants around and in the crevices of the two low outcrops of granite ledge that pop up in our north lawn, where we set up the croquet wickets in summer, but the low rocks and the plants—one near the final stake and one near the third wicket—just make our kind of casual croquet the more amusing. Many of my plants are not alpines or rock plants at all—simply what I happen to have handy. Thus far, I have not succeeded in having much color after June, when the moss pink and the miniature bulbs have stopped flowering. In early spring, the tiny wild tulips and wild daffodils are a joy, if they have managed to survive the cold winter and the depredations of the field mice. Allwood's Alpine Pinks bloom well into midsummer and bloom again if I cut them back, and Johnny-jump-ups continue to bloom until frost if I continue to cut them back and tend them—a big "if" nowadays. My plantings have grown too numerous to be well tended, now that labor is scarce and I have less time to work in them.

All gardeners who like to get down on their knees and grub enjoy reading books by others who do, especially if the authors write well and are good gardeners. Buckner Hollingsworth's new book, *Gardening on Main Street*, with drawings by Eva Cellini (Rutgers University Press), is, of necessity, as we shall see, not as substantial and solid a book as her earlier historical and biographical garden books (*Flower Chronicles* and *Her Garden Was Her Delight*), which were based primarily on research. Even so, the new book makes pleasant reading. Ten or fifteen years ago, the author and her artist husband moved from Rockland County, where she had struggled for twenty-five years with a garden on stony ground, to a house and lot seventy-five by two hundred feet, behind a low picket fence on the busy main street of a Vermont town. There were only seventy-six inches of weedy lawn between the house and the picket fence, and at the east side of the house a partially shaded spot of thirty by twenty feet. This problem area she made into her garden, with a small pool under an apple tree. The book is an entertaining blow-by-blow account of how, by trial and error, she turned this unpromising ground into a garden so full of bloom that it is now the pride of the community and a magnet for sightseers. She could have used her big back yard or part of the two-acre hayfield at the back of the lot and had the pleasures of privacy, but though she gives other reasons (drainage and soil and the fact that she was over sixty when she began), it seems to me that she really gardened where she did for the delectation of the town as much as for herself. Later, she turned the tiny front lawn into a garden with stone paths, and, finally, planted the sloping eight inches between the fence on Main Street and the sidewalk. She likes people; her neighbors and even the tourists wander in and out of her story. A lot of talk goes on over that fence.

Mrs. Hollingsworth is a scholar and researcher with a bookish turn of mind. Toward the end of her research for *Her Garden Was Her Delight*, she had a series of unsuccessful eye operations, and this ended for her all books of this sort—books based primarily on research. However, she barely mentions this misfortune and never complains or indulges in self-pity, nor does she say that reading must now come to her by ear. Luckily, she *remembers*, and her new book is full of literary and historical allusion as well as of practical and aesthetic advice. To me, it is a miracle that she can write at all now and can still garden.

I particularly relish the chapter on "Garden Gadgets and Gimmicks," and I plan to use her ingenious suggestion on how to turn those dry-cleaner's wire dress hangers that keep accumulating in every home into invisible plant supports (I never seem to have enough stakes or bamboo poles). I am with her on most of the flowers listed among her dislikes in the chapter on "Horticultural Horrors," but I take exception to her attack on Golden Glow. We had a fine stand of this hardy, carefree perennial when we bought our farm in Maine, many decades ago. It grew just outside our barnyard fence and was very gay against the gray shingles of the barn until the sheep, learning to lean through the fence, chewed it up. I replaced the plant with a prickly fire thorn that grew to be a handsome sight, but one winter our by then well-established *Pyracantha Kasan* was destroyed by field mice, who ringed the bark. Now I plant Burpee's huge Climax marigolds in the small square of soil, but they mean work and replanting each year. Golden Glow is an old-fashioned, leggy plant of somewhat inferior quality, but it seems *right* for early New England houses, and I enjoy the ones that flourish in our neighbors' yards. The clumps of yellow are cheerful as one drives along the highway, and are a time-honored spot of color in the New England scene.

My best news of the year is that Elizabeth Lawrence's distinguished first book, *A Southern Garden*, is again in print. I am lucky enough to own a first edition, now a collectors' item, which was published in 1942 by the University of North Carolina Press. This year the Press has brought out not a reprint but an updated illustrated edition that seems almost a new book. All the chapters of the old book have been kept, but much new material has been added. Miss Lawrence was living, writing, gardening, and practicing landscape architecture in Raleigh when she wrote the first version. She now lives in Charlotte, where she started a wholly new garden and began her excellent garden column for the Charlotte *Observer*. The new book adds a chapter of notes on the new garden, and photographs of her particular favorites among her shrubs and plants. Most charming of all the illustrations, though, is the frontispiece—an unintentionally romantic picture of young, pretty Miss Lawrence in her Raleigh garden, dressed in a flowing, long-skirted, short-sleeved summer dress, with her big dog beside her. The modest subtitle of both editions—"A Handbook for the Middle South"—is sound, but the book is far more than just a reference for gardeners in the mild climate of Zone 8; it is a literary yet scientific book by one of our best and most knowledgeable writers on gardens. Miss Lawrence writes with meticulous professional and scientific care, and, like the best English garden writers, she writes with humor and in simple, literary prose. To the new edition she has added a twenty-two-page chart of blooming dates—the earliest and the latest—of all the annuals, bulbs, perennials and biennials, shrubs, and vines she mentions, and a list (updated to September, 1967) of recommended nurseries that carry material for her part of the South. (These reliable sources of supply are situated all over the country, and therefore many of them carry plants for other climates.)

"I think of a garden," Miss Lawrence writes, "not as a manifestation of spring (like an Easter hat), nor as beds of flowers to be cut and brought into the house, but as a place to be in and enjoy every month of the year." This, of course, is not the way I can use a garden—in our Zone 4 to 5 strip of Maine coast, there are only a few days when winter garden living is possible—but I, a Northern amateur, come lately to any real knowledge of what gardening is about, have learned more about horticulture, plants, and garden history and literature from Elizabeth Lawrence than from any other one person. *A Southern Garden* is far more than a regional book; it is civilized literature by a writer with a pure and lively style and a deep sense of beauty.

The Art of Flower Arrangement, by Beverley Nichols, a luxurious Viking Studio Book, was published in the last month of 1967, but it is so beautiful and interesting that I mention it for any who have not seen its seventy-odd brilliant color plates, most of them reproductions of ancient paintings, illuminated manuscripts, early flower paintings, etc., and, to wind up Mr. Nichols's historical narrative, its excellent color photographs of early-modern to present-day flower arrangements. He is a prolific writer on many subjects and has written many lighter books on garden themes—mostly drawn from experience. In the garden-book field, this volume is his magnum opus. The text pleases me to an extent, I must admit, because his opinions coincide with mine, with one big exception: he seems to blame the low estate of modern flower arrangement, at least in part, on the modernist French and European painters. (See the chapter headed "Chaos.") I think that any good artist has much to teach the flower arranger, especially on the matter of simplicity, even if the flowers he paints are not like any flower on earth. The author, who is also a lecturer and authority in garden circles in many parts of the world, has apparently run afoul in this country of some of the absurdities created by the rigid rules for the "standard" flower shows that are put on by the Federated Garden Clubs. As he points out, flower arranging is now Big Business, and he doubts that an art—or sub-art—can be mass-produced. I agree. His big book covers more than flower arranging—such interesting subjects as ceramic artificial flowers (which are often delightful, especially when they do not pretend to be natural) and the miniaturists, starting with Fabergé's exquisite flower jewelry. He also pays proper tribute to the influence of Constance Spry in England and this country as a semi-modern arranger.

Another new big de-luxe volume of this year, on a highly specialized subject, is *Table Decoration: Yesterday, Today & Tomorrow*, by Georgiana Reynolds Smith (Tuttle), which is outstanding for its hundred and seventy illustrations, a few of them color photographs of modern flower arrangements by the author. The most interesting are the black-and-whites, taken from ancient engravings and line drawings and photographs of objects used as table decoration (or pictures showing the lack of it) from the distant past to the present. Mrs. Smith, better known to the garden world as Mrs. Anson Howe Smith, is a lecturer and a gifted flower arranger, and she arranged most of the period flower bouquets used to illustrate Margaret Fairbanks Marcus's enduring *Period Flower Arrangement*. Mrs. Smith has been a lifelong student of design, and for many years she has been doing research and making card-catalogue notes on table decoration and table habits over the ages. It was Mrs. Marcus who encouraged her to turn those notes into this book. Mrs. Smith is less well organized, as a writer, than Mrs. Marcus, partly because instead of writing a progressive historical narrative she has chosen to write on themes, and has a tendency to hop about from century to century within the theme of each chapter. Take the chapter on "Garlands, Chaplets, and Strewn Flowers." You find yourself first in Melanesia with a Polish anthropologist in 1922 (garlanded canoes). In a trice you are in ancient Greece and Rome; then, after passing through the Dark Ages, the Middle Ages, and the Renaissance, you surprisingly soon are in Cairo, under the Khedive, in the eighteen-nineties, at the garlanded banquet table of a party given in honor of our consul, until you wind up, breathless, in the present and a discussion of contemporary Christmas-table decorations. Six pages further on—in the block of illustrations that separate this chapter from the next and illustrate both of them—you are attending, thanks to the English painter William Salter, one of the famous annual Waterloo banquets at the Duke of Wellington's London residence. The ornamental centerpiece is a silver-and-gilt surtout presented to the hero of Waterloo by a grateful Portugal. Not a garland is in sight in the photograph of the big surtout, but the author points out that the hands of the carved nymphs who dance around the surtout are outstretched to *hold* garlands. In Salter's painting, the survivors of Waterloo have grown old and portly; ribbons and decorations cover their expansive chests. After you get used to the jumpiness, the book is a lot of fun, for the author avoids the well known, in both art and fact, and leads one into new territory.

Mrs. Smith has, for example, taught me a new word—"nef," a noun from the Middle Ages, though still in Webster's Unabridged. A nef was a golden or silver table piece, usually in the form of a ship, used to hold utensils or merely as an ornamental centerpiece. (Table decoration was by no means always flowers or fruits. It involved non-floral ornament from the very early days, and drew on the skills of the silversmith and the goldsmith, the maker of fine porcelains, the glassblower, even the confectioner.) One full-page plate shows the famous Burghley Nef; the ship is a nautilus shell, mounted in silver and resting on a silver mermaid's back. Another nef, made in Germany in the sixteenth century, is mechanized; set in motion, its sailors manipulate the rigging of a three-masted ship, and on the deck a band plays and courtiers bow before Emperor Charles V. All in all, this is an unusual book, and a beautiful one. (It was designed and printed in Japan.) It contains much social history and much knowledge of eating habits through the centuries and throughout the world. The captions for the illustrations are informative and were probably once card-catalogue notes. I especially liked this one: "Herbaceous flowers '*en chemise*,' recommended in an early 19th century cook-book. The flowers first had to be dipped in white of egg and then rolled lightly in powdered sugar to give a hard, glistening surface resembling the fashionable porcelain flowers." These real flowers—*en chemise*, I take it—were for those who could not afford the then fashionable porcelain ones. They *look* like porcelain in the color photograph, for which the author must have dipped the flowers herself before she arranged them. It gives me ideas for next summer. Chemised cosmos, rose, larkspur, verbena, nicotiana, and salpiglossis might be less irritating to a husband who suffers from pollinosis than the naked flowers. Certainly they would be more durable.

14

Knots and Arbours—and Books

Elecampane. From Pierandrea Mattioli,
Commentaries on the Six Books of Dioscorides, *1563*

March 28, 1970

A certain amount of snow and gloom of night managed to stay this courier from the swift completion of her December rounds, but March is a good time to talk and think about the world of the garden. By March, for those of us who live in the Northeast, the summer seed and plant orders are in. From Washington north to the Canadian border and east to Maine, the tender seedlings and plants raised in hotbeds, cold frames, or greenhouses must wait for that final snowstorm before being put into the ground. The gardener has finished his mid-winter reading of Christmas gift books and laid his plans for new enterprises for the coming summer. It is time for him and for me to get out of our armchairs and take stock. My own reading of the ever-spreading flood of books related to gardens, flowers, and horticulture goes on throughout the year, and this winter it brought me up short with the question of why more of the books I happened to enjoy are of English origin. With a few shining exceptions, American "garden writers" are a drab lot. We publish each year an enormous number of books related to the garden, but except for our excellent reference books they are mostly how-to-do-it books offering advice—sometimes bad advice, sometimes good—or they are superficial and overexpensive de-luxe gift books of one sort or another that are bought for their color illustrations, their huge size, and their display value rather than for their text.

Unlike this country, England has since the Elizabethan Age regarded books that deal with horticulture and plants as a true branch of literature, and on that account has been rewarded by a long line of writers and illustrators, from the ancient herbalists on, whose books give pleasure not only to gardeners but to any other reader. The good books got written as early as the sixteenth century, when John Gerard's *The Herball, or Generall Historie of Plantes* was published in London in 1597, to be followed by John Parkinson's *Paradisi in Sole Paradisus Terrestris* in 1629. These two primitive doctors, of course, intended their books only as medical manuals, but both men raised their own herbs and flowers, and they are remembered less as early botanists and doctors than as the first of the writers whose books on the lyrical delights of the garden became a part of English literature and set a standard that English writers have tried to live up to. The same has been true of English botanical artists.

In spite of its then fashionable punning Latin title, the Parkinson book is written in English prose. The translation spells out the play on words: "Paradisi in Sole" (Park in Sun's), "Paradisus Terrestris" (Earthly Paradise). It is the book that centuries later engendered Juliana Horatia Ewing's sentimental children's book *Mary's Meadow*. The Ewing book reflects the staying power and charm of "Earthly Paradise," but Mrs. Ewing and her era, not Parkinson, must be held responsible for its sentimentality and moralistic overtones. Parkinson's was a far purer and more genuine delight in the natural world. The gardens of Elizabethan England were "physic" herb gardens, planted in elaborately patterned "knots and arbours." John Gerard was a "Master of Chirurgerie" and Warden of the Company of Barber-Surgeons; Parkinson was apothecary to the first King James and held the position of botanist under King Charles. They live on in history, however, as the first of the great writers in English who celebrated the importance of the garden, the flower, the plant.

There were other English herbalists, of course, writing both before and after Gerard and Parkinson and contemporaneously, and even the Celts are said to have had a primitive herbal, but the real golden era of the writers on horticulture covered the reigns of Queen Elizabeth and James I, and their writing was the purest of Elizabethan prose. Nicholas Culpeper, a near contemporary, is often ranked as a writer of high importance, but because his herbal was governed by astrology, magic, and myth and because he was generally, I gather, a sourpuss in his first herbal and gave no credit to anyone since Galen, I shall not include him. For careful estimates of the virtues of Culpeper, Gerard, and Parkinson, I refer you to the last book in my list below—*Early American Gardens*, "*For Meate or Medicine*," by Ann Leighton.

It surprised me to discover that even in the sixteenth and seventeenth centuries any new book aroused great controversies and charges of plagiarism. Gerard and Parkinson did not escape. In fact, they were in the thick of it, and the shenanigans of their publishers sound astonishingly like those of our own publishing houses, which, if one type of book is successful for a rival house, will immediately try to duplicate it with books of their own. Gerard, especially, was in a pickle. Claims were made, and are still made (cf. the Encyclopædia Britannica), that his herbal was lifted from one written in Flemish by Rembert Dodoens in 1554. I hope the editors of the newest edition of the Britannica will reread Gerard and study the evidence developed since their dismissal of him, forty years ago, in the fourteenth edition, which is the one I happen to own.

John Gerard was born in 1545 and died in 1612. He was fifty-two when his first and only book was published, and he was a practicing surgeon and a highly qualified general medical man and botanist. He also superintended the large gardens of his patron, Lord Burghley, and tended his own little plot, to which he gave "especiall care and husbandry," in Holborn, where he was a neighbor of Shakespeare. For his own garden he assembled rare plants from fellow-gardeners and from friends who were plant collectors, and in his herbal he gives generous acknowledgment to each individual who helped him put together his own rare physic garden. Certainly Gerard did *borrow* from Dodoens' herbal, *Pemptades*, which had been the chief medical reference book in Europe and England since its publication, but he probably felt no compulsion to admit the debt. He does admit his debt to others in his dedication of his book to Lord Burghley, saying that he has worked for twenty years on "the large and singular furniture of this noble Island" and that he has added "from forreine places all the varietie of herbes and floures that I might in any way obtaine." But Dodoens in turn had lifted from earlier German herbals and from the Greek and Roman herbals. And Gerard certainly wasn't borrowing from anybody when he wrote the following account of elecampane, quoted in Miss Leighton's book: "It groweth in meadows that are fat and fruitful; it is oftentimes found upon mountains, shadowie places that be not altogether drie; it groweth plentifully in the fields on the left hand as you go from Dunstable to Puddle Hill; also in an orchard as you go from Colbrook to Ditton ferry, which is the way to Windsor, and in sundry other places, as at Lidde, and Folkstone, neere to Dover by the sea side." He doubtless made his own "Generall Historie of Plantes" from his own garden of rarities and his plant collecting with the help of neighbors and friends. He enlarged this list from Dodoens' European list and other sources. Miss Leighton points out the amusing fact that if Gerard borrowed from Dodoens, John Milton in turn was a borrower from Gerard. When Adam is searching for

> *Some tree whose broad smooth leaves together sewed*
> *And girded on our loyne, may cover round*
> *Those middle parts*

he and Eve go into the woods, "where soon they chose/The Fig tree." The fig tree is "not that kind for Fruit renowned," and Milton's description of it comes straight from Gerard's description of the fruitless "Arched Indian Fig Tree."

The second edition of Gerard's herbal, now commonly known as *Johnson-Upon-Gerard*, was published in 1633, after Gerard's death. It was edited, corrected, and amplified by Thomas Johnson, an apothecary and thus a well-qualified botanist and herb man. At first, Johnson sounded to me like an opportunist, but the truth is that Gerard's original herbal was full of errors of identification, particularly in the illustrations. Without the assist from Thomas Johnson, Gerard's only book, so full of quotable, good writing, might never have survived. In one year's time and in a hurry, Johnson revised Gerard's herbal in order to have it published before the work of Parkinson's old age—a ponderous medical book titled *Theatrum Botanicum, An Herbal of Large Extent.* Johnson's life was short. He was a Royalist and died of a wound while fighting for the King against Cromwell. Ten copies of *Johnson-Upon-Gerard* arrived in America with the early New England settlers.

It must not be forgotten that these men lived in one of those extraordinary periods in history when men of genius and wisdom are men who can *write.* A parallel could be made with the birth of this republic. The words of the Declaration of Independence ring in our ears today. The words of those gifted Americans who, after long, slow journeys, arrived in Philadelphia in 1787 to draw up a constitution for the newborn republic—these words are our most precious heritage. Written, not spoken, language is what counts and what endures. A closer parallel could be made with the period of Dioscorides, whose herbal, *De Materia Medica*, written in 50 A.D., was the only medical botanical reference book available to doctors during the Dark Ages. There were other writers in ancient history who studied plants and nature and wrote herbals. Theophrastus, a pupil of Aristotle, wrote his *Enquiry into Plants* in the great days of Athens, three centuries before Christ; the elder Pliny wrote an herbal in the first century A.D., and the younger Pliny's letter on the delights of the natural world is one of the great pieces of nature writing that have survived. But for this first golden era of the herbal I shall instead commend to you Miss Leighton's introductory chapters, Chapter 1 of Buckner Hollingsworth's *Flower Chronicles*, on herbals, and her valuable bibliography for this book.

The three herbals by Gerard, Parkinson, and Culpeper had three different illustrators. Most of the plant pictures in Gerard's are from wood-block prints now generally attributed to a German artist named Bergzabern, who signed them with a Latinized version of his German name, Tabernaemontanus—a clumsy signature for an artist if ever there was one. His pictures are delightful—by far the best, as art, of the illustrations for the three herbals. They add greatly to Gerard's felicitous text. The illustrations for Parkinson are insignificant as art and oversimplified botanically, but are more helpful for the identification of plants than Bergzabern's. Culpeper's illustrations are the most realistic of all and are pleasant to look at but not distinguished. All herbalists seem to have borrowed their illustrations from earlier herbals. How many still earlier borrowings there were along the line it is impossible to know. The majority of the illustrations of plants in the Leighton book are taken from the Gerard herbal, and you can see why they enhanced Gerard's mixture of medicine, gardening, and joy. Gerard is universally credited with being the first herbalist ever to publish a picture of a potato. A portrait of the author in his book shows him in what I suppose was the formal dress of an Elizabethan doctor. He wears a ruff, puffed sleeves, a pleated waistcoat, and in his hand he holds a specimen of a Virginia potato plant.

Gerard lifts from Dodoens. Milton lifts from Gerard.

Bergzabern lifts from Fuchs. I, at this late date, make use of the research done by Mrs. Hollingsworth and Miss Leighton and others. What are books for, anyway, if not to be absorbed and used?

Among the garden writers of the Elizabethan Age, Francis Bacon, a contemporary of Gerard and Parkinson, cannot be omitted, if only because his forty-sixth essay, "Of Gardens," must have been far more widely read in England when James was on the throne than any of the new herbals. It and his preceding essay, "Of Building," a companion piece, were addressed to the King and were doubtless written to advance the author's political ambitions, which were doing better under James than under Elizabeth. The two essays describe a grandiloquent dream garden and palace possible only for a monarch of limitless wealth. Nevertheless, it does come through strongly that Bacon, like almost every other Englishman, knew and loved gardens and must have had one of his own: "God Almighty first planted a garden. And, indeed, it is the purest of human pleasures. It is the greatest refreshment to the spirits of man, without which buildings and palaces are but gross handyworks." He goes on to describe a garden of thirty acres divided into three parts; the central plot, a mere twelve acres, was to be pleasant to the eye every month of the English year. Fountains, lawns, trees, alleys, ornaments, vistas, turrets, and towers, but always, too, the humble flowers of a home garden Bacon himself might havé grown. Under "April" he lists "the double white violet, the wall-flower, the stock gilliflower, the cowslip, flower-de-lices, and lilies of all natures, rosemary flowers, the tulip, the double peony, the pale daffodil, the French honeysuckle, the cherry tree in blossom," and other blossoming fruit trees and the lilac. The essay on gardens is not one of Sir Francis's best efforts, and certainly not one of his better-known or most-quoted essays. Although he was a garden lover and a moralist, he was far from admirable. He spent the last five years of his life in disgrace, after twenty-three charges of corruption were brought against him. He was committed to the Tower, only to be released a few days later by the king he had served so well. During Elizabeth's reign, he was out of favor, and his earlier patron, Lord Essex, lost his head—by order of the Queen. But it might soften our memory of a queen who ruthlessly beheaded her enemies if we remember that Elizabeth, in the autumn of 1555, before she came to the throne, went down to Hatfield and planted a row of trees, some of which were still going in the early years of this century.

Bacon, in his essay titled "Of Studies," which is the first chapter in the first edition of his *Essays*, published in 1597, wrote, "Reading maketh a full man," but he also said, "Some bookes are to bee tasted, others to bee swallowed, and some few to bee chewed and disgested." A conscientious book reviewer must actually read all the new books, but most of us can follow Sir Francis's sensible advice. I regret that in the list of recent garden books that follows only a few fall into the chew-and-disgest category:

Wild Flowers of the United States, *Volume 3*, *Texas*, by Harold W. Rickett, in Parts 1 and 2, boxed, of the New York Botanical Garden's in-progress series, edited by William C. Steere, with color photographs, line drawings, and diagrams (McGraw-Hill). This volume of this massive enterprise —to photograph, identify, classify botanically, and describe for easy identification by amateurs the wild flowers of the United States—is the most surprising and probably the most valuable to botanists and amateurs alike of the ones thus far published. Texas, because of its size, its highly varied climates and soils, and its influx of Mexican tropical flora along the southern border, has more species of wild flowers than any other state. It has, however, been laggard in the scientific recording of its flora. Thus, some of the flowers described in Dr. Rickett's admirable text, and reproduced in excellent color, appear for the first time in a general reference book. Sensibly, he has not followed strict taxonomic rules for his grouping of plants under genera and species but has grouped them for ease of identification. The Million-Dollar Book, which is what Volume 1 was called, as the first of the series designed to cover the entire country, is now on its way to becoming a multi-million-dollar enterprise. Texas money helped finance this third volume, and if this important series is to be completed, other states and other benefactors will have to help. The color photographs are the best yet in the series. Cactuses, which we think of when we think of Texas, are—except for a few small ones—excluded, because they are flowering woody plants, not wild flowers, but look at the beautiful color page of Yuccas! Best of all are the exquisite photographs of delicate flowers a non-Texan has never seen the likes of.

Wild Flowers of Connecticut, by Dr. John E. Klimas, Jr. (published for the State of Connecticut Audubon Society by Walker), and *Wild Flowers of North Carolina*, by William S. Justice and C. Ritchie Bell (University of North Carolina Press). Local field guides of wild flowers are an important adjunct to the comprehensive tomes like Harold Rickett's *Texas*, if only because by the time one brings home an unknown flower it is usually shattered beyond recognition and identification. One by one, the states are bringing out state field guides of their native species and of flowers that have escaped from gardens and gone wild. Pocket guides that are even more limited and local proliferate. There is a great difference in the quality and usefulness of these small books, as these two will show. *Wild Flowers of Connecticut* is a pretty little pocket piece, but its title should have been "Wild

Flowers to Be Found in the Larsen Audubon Sanctuary, Fairfield, Connecticut." The ninety-six flowers listed were all studied and photographed within the hundred and ten acres of the Sanctuary. Connecticut is not a big state, but its terrain is varied—hills, bluffs, fertile river valleys, and coast-line. The wild flowers listed in the little book are by no means representative of the state, and the list omits many of the commonest and showiest of Connecticut's wild flowers. The guide is organized merely to aid the casual visitor to the Sanctuary and to engage his interest in wild flowers as well as in birds. The flowers are, sensibly, grouped by color and by season, but the color photographs, two inches square, are far too small for ease of identification and, worse, are con-fusing because too many of them show only details of a flower instead of the whole flower stalk against its natural back-ground. (I was even fooled into thinking that I had come on a brand-new flower by the handsome color photograph of a small section of the inflorescence of a cardinal flower.) The illustrations are separated from the descriptive text, which is always an annoyance. All this is too bad, for the color reproduction is good (no acid greens) and Dr. Klimas's commentaries and descriptions are top-notch—well written, and just that relaxed blend of science, myth, legend, and odd fact that is right for the average amateur like me. A state-wide Connecticut guide, though, has yet to be published. *Wild Flowers of North Carolina* is a heavier, bigger book, but not so big or heavy as to be a burden in the capacious pockets of a hunting jacket. The book could be a model for any state field guide. The photographer, Dr. Justice, is given top billing, and his four hundred remarkable full-color photo-graphs set the format of the book. They range from three-quarters of a page down to two-by-three-inch pictures not only of wild flowers but of wild flowering vines, shrubs, and trees. All the plants are growing in their natural habitat, and most of the pictures are of the whole plant rather than of a detail, except when the detail is the chief point in quick identification. Right next to its picture by Dr. Justice is its description by Dr. C. Ritchie Bell, of the University of North Carolina, whose prose is non-technical but does not talk down. Dr. Justice, a surgeon, has found time to become an excellent color photographer and field botanist. If there is a fault in the book, it is in the color reproduction, for some of the woodland greens are too blue. There are a good introduction, a brief, elementary botanical foreword, charts, diagrams, a map, and so on—everything a student or an amateur needs to know. North Carolina has its mountains, piedmont, and coastal plain, and this book is comprehensive enough to be useful in much of the Central Atlantic area, which covers eight other states, from Delaware to Georgia and from the coast to Kentucky.

The World of the Japanese Garden, from Chinese Origins to Modern Landscape Art, by Loraine Kuck, with forty-four color photographs by Takeji Iwamiya, and one hundred and fifty-six black-and-white photographic plates (Walker). This unusual book is my nomination for the most beautiful of the big, expensive garden books listed here. Takeji Iwamiya is Japan's leading color photographer, and his color plates are bright and airy, or cool and dark, and all of them are haunting. (This is a book about Japanese gardens, ancient and modern —*not* about Japanese flower arrangement.) The Japanese artificial garden of raked sand, pebbles, stones, pools, and flowering shrubs has its symbolism for the Japanese, but in our small cities and the crowded suburbs of our big cities it has proved to be practical as a minimum-upkeep substitute for a lawn, and any attempt to introduce the symbolism of Japanese *gardens* can be forgotten. On the other hand, only a scattering of homes in the United States make good backgrounds for symbolic Japanese flower *arrangements*, which lose their meaning without the ritual and symbolism of each carefully placed leaf or flower and the hallowed niche in which they are displayed. Miss Kuck's readable text traces the Japanese garden back to its Chinese origins and becomes, indirectly, a history of Japan itself. Printed in Japan, the book in format and illustration makes some of the big picture books printed in England or France, with their usual excellent color plates, look almost heavy-handed.

The Color Dictionary of Flowers and Plants, for Home and Garden, by Roy Hay and Patrick M. Synge (published by Crown in collaboration with the Royal Horticultural Society). A color dictionary of flowers and plants is a new idea, and this big volume, published first in England, was the handsomest garden book of 1969. Two thousand and forty-eight plants, including some trees and shrubs, are shown in color photographs two and a half inches square, eight of them centered on a page, leaving wide margins for Latin names, English variety and familiar names, and numbers that key each picture to the dictionary section of the book, where the plant or tree is briefly described. In addition, there is an excellent section of cultural notes. The photographs sparkle and glow and are a delight to study. Yet with all this comprehensiveness, care, and beauty, an American gardener lusting to buy, say, the seed of "*Dianthus*, 'Warrior' (Border)"—the last word meaning that it is a hardy border pink—or a balled-and-burlapped and ready-to-set-in-the-ground Silver Fir tree with blue-mauve cones (*Abies forrestii*) would be hard put to it to know where to make his purchase. This gardener, who collects catalogues and grows her garden pinks from English seed, has never happened to see the seed of *Dianthus*, "Warrior" (Border) or the Silver Fir tree with blue cones in any American or English catalogue. Britons may know where to obtain these treasures; what the book lacks, for American readers, is a list of sources of supply, in both England and the United States. George Kalmbacher, taxonomist of the Brooklyn Botanic Garden, was the consultant for the American edition, and in his foreword he hopes that this work will accelerate the "international interchange of materials." His only advice to Americans is to ask their County Agent or State Agricultural Experiment Station whether the plant they want is suitable for the climate zone in which they live. But where to *send* for a plant, tree, or seed? Well, buy the book and have a good time dreaming. Many of the color photographs do carry names common to both countries, especially in the Dutch-bulb section, but many do not.

Shakespeare's Flowers, by Jessica Kerr, with color illustrations by Anne Ophelia Dowden (Crowell). The author is Irish, and the illustrator, although American, might just as well be English as far as the style of her flower portraits goes. Her luminous and precise water colors, near to perfect in reproduction, remind one of Graham Thomas's delightful water-color illustrations for his own enduring books, or the color plates in *The Oxford Book of Wild Flowers*, except that they have more polish. The publishers label the book for "ages 12 and up." It would be a rare American twelve-year-old who knows the Shakespeare songs and plays or the differences between the English and American familiar flower names well enough to read the book without help. It is, however, a perfect book to read aloud to a bookish, flower-loving child by someone who could help transfer Shakespeare's "cowslip bells," "lady-smocks," and "cuckoo-buds" first into the current English familiar names and next into the American equivalents. This slender volume is the best of the many books on Shakespeare's flowers I have encountered. Most of the others are either scholarly and dull or sentimental and amateurish. The book is backed by well-concealed scholarly research by both author and illustrator. The prose is relaxed and entertaining, and the water-color plates are delicious. All in all, an exquisite brief book for adults, with a bibliography, helpful indexes of Latin and British familiar plant names, and scene and line references to the plant passages in the plays and poems. Shakespeare was equally fond of garden flowers, wildings, weeds, herbs, and trees, all of which are covered here in text and picture. Too bad the American edition could not have added a glossary to translate the English familiar names of marigold into our calendula, cowslips into primroses, and so on, and to explain our own confusing regional differences in familiar plant names.

Flower Decoration in European Homes, by Laurence Buffet-Challié, with forty-one color plates and eighty-seven black-and-white photographs of flower arrangements done by Jacques Bédat (Morrow). Did you know that a flower arrangement showing a "Sylvania Sunlamp with flowers and fruit expresses peace and contentment in the home"? Neither did I. A review in a Southern newspaper of a book titled *Interpretive Flower Arrangements* assures us that this is so, and says that the author, Nelda H. Brandenburger, gives us "easy directions for expressing any emotions" and tells "how to win prizes on many subjects." Gardeners who have no interest in winning prizes but take pleasure in bringing flowers into their homes and putting them into vases, whether carelessly or carefully, and who are disenchanted with the current books on flower arranging that contain just such grotesqueries as the Sylvania Sunlamp may enjoy Buffet-Challié's expensive book of so-called traditional European arrangements. In Europe and Scandinavia, it seems, bouquets are for decoration of the home and do not try to interpret a theme or be symbolic. How refreshing! The book is of a size and a format that classify it as one of the "big beautifuls," or gift books, but in this case the color photographs and the arrangements themselves lack charm and originality, and few of them are actually shown in people's houses. M. Bédat's style is too modern and "arranged" to be "traditional." The American title of the book leads one to expect traditional arrangements in homes by many European Constance Sprys. It is the text that matters here, once the author has finished with how to put flowers in a vase with all the modern props and gadgets.

Flowers in the Garden, by Dorothy Jacob (Taplinger). Mrs. Jacob, the author of *A Witch's Guide to Gardening*, which is a lively brew of legend, myth, folklore, and literary scholarship, has now written a very different sort of book—her account of seventy years of down-to-earth planning and growing of gardens, intermingled with memories of her family and reflections on the glories and sorrows of her native England. She divides the book into four sections: "Spring1889–1914," "Summer 1915–1937," "Autumn 1938–1958," "Winter 1959–," which last means to 1968, when the book was first published, in England. While it is not one of those literary masterpieces on gardening that British writers quite often used to achieve, the book is so sincere, so personal, and so full of the observations of a practical yet reflective gardener that no one who feels that gardens and trees and home landscaping go far beyond the plants involved will want to miss it. The First World War and postwar chapters are particularly poignant, as are the author's struggles to garden elaborately without adequate help when she was a widow in what she calls Heartbreak Hall, a castellated Tudor mansion with thirteen acres of garden and park, and then in Comfort Corner, a farmhouse with only a half acre of garden. I cannot always agree with her ideas on landscaping, especially as they are shown in the dim photographic illustrations, or even with some of her gardening advice (extreme neatness and no casualness), or with her taste in planting flowerbeds, but I do share her dislike for dahlias and her interest in garden history and her bad gardening habit of overexpansion. The book is best summed up by saying that I can't think of a pleasanter circumstance than to have the practical and thoughtful Mrs. Jacob, now in her eighties, as a next-door neighbor and friend.

V. Sackville-West's Garden Book (Atheneum). If you want a garden book that is notable for the verve and high style of its writing and for its near-to-professional knowledge of botany and horticulture, turn to this anthology, unless you already own the four collections of garden notes drawn from the author's weekly garden columns in the London *Observer* between 1947 and 1961. The book is an anthology of the four previous anthologies, the very idea of which filled me with misgiving. Oddly, it works. The selections have been astutely chosen by Philippa Nicolson, the author's daughter-in-law, and arranged not in chronological order but by chapters for each month of the year. (All good gardeners are forever remembering past errors or dreaming ahead to next year's plantings.) Miss Sackville-West grew up in the great landscaped acres of Knole, but she had no interest in gardens until she married Harold Nicolson, so in this book he is a presence behind the scene. He was responsible for the layout of paths, hedges, and walls, and he was a partner in all her gardens. Her highly unconventional gardening ways—first, and briefly, at Long Barn, and for the rest of her life at Sissinghurst Castle—are fascinating. Let not the "stately homes" alarm you. Mrs. Nicolson was the least snobbish and least careful of gardeners. Her sensible garden experiments give the average amateur a new idea per page, but the book can be read for the pure pleasure of its prose style. Her daughter-in-law has said it best: "She was a romantic, and yet she was very practical; she was experimental, and yet she respected tradition; she was literary, but she wrote from knowledge, not from books. . . . The first principle was ruthlessness. You must never retain for a second year what displeased you in the first. . . . She was the opponent of too much tidiness." There should be more gardeners like her—eccentric, didactic, ruthless. The last of her notes were written a year before her death, in 1962, and no one has yet appeared to take her place. Beautiful photographic illustrations in both color and black-and-white.

Common or Garden, by Tyler Whittle (Heinemann, London). In his earlier *Some Ancient Gentlemen*, the author, a knowledgeable and often witty English broadcaster and garden writer with a literary turn of mind, gave promise of carrying on from where Geoffrey Taylor, another Englishman, left off, in 1952, with those two enduring, charming, and learned but non-pedantic little books—*The Victorian Flower Garden* and *Some Nineteenth-Century Gardeners*. Mr. Whittle's new book does not fulfill that promise. It is less unified than *Some Ancient Gentlemen*, and there is more sense of effort in his humor and some repetition of ideas, themes, and stories—inevitable for broadcasts, perhaps, but not necessary in books. It is, even though a bit flibbertigibbety, far more readable than the standard American garden book. Mr. Whittle takes one all over the map of Europe—sometimes as plant explorer, sometimes as tourist, sometimes as renter of Italian villas. The book is replete with advice, horticultural history, bright ideas, and entertaining anecdotes. He is especially amusing on the lunacy of fanatic gardeners. His anecdote about Walter Savage Landor is just one example. Landor, enraged by an indifferent dinner served by his cook in an upstairs dining room in Florence, threw the cook out the second-story window. Struck by horror at his act, but with never a thought of the man writhing on the ground with broken bones, Landor cried, "My God! I had forgotten the violets!" Mr. Whittle writes of the United States, too, and his facts and names are often inaccurate. Wallace Stevens was *not* a "Pennsylvanian poet." We do *not* visit "The New York Botanic Garden" in the Bronx. And so on. Contrasting the "Great Glass Houses" of Victorian England and the conservatories of modern days, he mentions a visit to the huge, ultra-modern Shaw Climatron, in St. Louis, and cites as "rarely seen plants" that he viewed there the Hong Kong Orchid Tree, the Sea Grape, and the Red Powder Puff. All three are common sights to anyone who has ever lived in Florida. The Sea Grape, a native of Florida's sandy shores and bayous, is beautiful. The non-native Orchid Tree is a handsome ornamental lawn tree all too common in Florida dooryards. So is the Powder Puff, but to me it is more interesting than attractive as an ornamental shrub—messy, straggly, and forever littering the ground with its untidy faded brown puffs. Let us hope that Mr. Whittle's entertaining account of the Duke of Devonshire and Joseph Paxton's Grand Conservatory at Chatsworth is more factual. It surely must be, for Mr. Whittle's is just a merry and up-to-date version of an oft-told tale—particularly *well* told in Geoffrey Taylor's *The Victorian Flower Garden.* The Grand Conservatory, or Great Glass House, which covered an acre of ground, belonged to William Spencer Cavendish, sixth Duke of Devonshire, an enormously wealthy bachelor. Joseph Paxton, his head gardener and intimate friend, was the architect of the huge glass building, with its miles of pipes. He was second only to John C. Loudon as a scientific gardener of the first half of the nineteenth century, but his success as architect of glass buildings led him away from gardening and on to the construction of the Crystal Palace, built for the London Exhibition of 1851, for which he was knighted by Queen Victoria. Mr. Whittle brings us up to date on the destruction of the conservatory, but Paxton lives on as an early visionary who led us on toward the fantastic Climatron in St. Louis and even, perhaps, as the architect who steered us toward our chilling glass-walled skyscraper cities.

Early American Gardens, "For Meate or Medicine," by Ann Leighton, with eighty-four illustrations (Houghton Mifflin). This valuable, wide-ranging book is both a history and a documentary. Miss Leighton writes well: the first two sentences of the book read, "The gardens of any period in history are its most intimate spirit, as immediate as its breath, and as transient. Yet, unlike all else about a particular time, they are capable of being recaptured and re-created today, in essence and in fact." Gardeners interested in the history of gardens, especially those who like to follow the travels of seeds and plants from one part of the world to another, will all want this odd, unusual book. It will be of equal interest to the non-gardener who is searching for information about our earliest New England forebears. The Pilgrims and the Puritans came prepared to grow their food and medicine and brought with them for guidance the Elizabethan herbals, including those of John Gerard, John Parkinson, and Nicholas Culpeper, as well as Thomas Johnson's *Johnson-Upon-Gerard* and Rembert Dodoens' *Pemptades*. Miss Leighton is by avocation an antiquarian gardener. She plants and tends a period garden for the restored 1640 Whipple House, in Ipswich, Massachusetts. She is at heart, though, a historian, antiquarian, and demon researcher. The eighty-odd illustrations she has dug up show her range of research, and in themselves they are worth the price of admission. There are maps, diagrams for geometrical "knots and arbours," title pages of ancient books, illustrations from the old herbals, and such items as a double-page facsimile of Cotton Mather's list of the plants of the New World that he felt the Royal Society in London should know about because he had not been able to find them in any other herbal. The last half of the book Miss Leighton calls an appendix. In it she lists two hundred of the earliest settlers' commonest plants, names those who grew them, and records what they expected of each plant. The wild differences of opinion and the astounding variation of the evils and virtues ascribed to each plant make hilarious reading. I have omitted any mention of what she has to say about "Meate," although the book is full of recipes and references to the uses of plants for food and for the still-room; it seems fairer to leave this part of the volume to the experts in gastronomy. There are many recipes I would like to try myself. I could wish that Miss Leighton had used fewer quoted antique words, fewer words with antiquated spellings, if only because they make the book too full of quotation marks, but then we perhaps would not have had the quotations from Gerard and Parkinson, which are delightful.

Editor's Note

There is a chapter missing from this book. Katharine had planned to write a final piece—about the gardens of her childhood. Because of her failing health, the last chapter never got written. This photograph is all that remains, to substitute for it. Here we see her Aunt Kitty (Catharine De Forest Sergeant) at the end of the garden path at 82 Bridge Street, Northampton, with black attendant in the foreground. The flowers at the left just might be cosmos.

Seedsmen and Nurserymen

Allwood Brothers Ltd. Hayward's Heath, Sussex, England. *(page 11*)*

Armstrong Nurseries. 1265 South Palmetto Avenue, Ontario, Calif. 91761. *(page 76)*

Barnhaven Gardens. Formerly Gresham, Ore. Taken over by Jared Sinclair, England. *(page 39)*

Bay State Nurseries. Became Seaside Planting, Halifax, Mass. *Out of business in 1966. (page 106)*

Bobbink & Atkins. Now Bobbink Nurseries, Inc., P.O. Box 124, Freehold, N.J. 07728. *(page 9)*

Bosley Nursery. Mentor, Ohio 44060. Wholesale growers only. *(page 34)*

Breck's Seeds. 6523 N. Galena Road, Peoria, Ill. 61632. *(page 13)*

Bristol Nurseries. Bristol, Conn. 06010. *(page 150)*

Brownell Farms. P.O. Box 5965, Milwaukee, Ore. 97222. *(page 110)*

Brownell Roses. Markets hybridized roses through Stearn's Nursery, Geneva, N.Y. 14456, and Field Nursery, Shenandoah, Iowa 51601. Herbert Brownell continues to hybridize stock in Little Compton, R.I. 02837. *(page 82)*

Buell's. Eastford, Conn. 06242. *(page 60)*

W. Atlee Burpee. Now Burpee Seed Company, Warminster, Penn. 18974; Clinton, Iowa 52732; and Riverside, Calif. 92502. *(page 3)*

Champlain View Gardens. Now owned by de Jager Bulbs, Inc. (*see* below). *(page 149)*

Cherry Hill Nurseries. W. Newbury, Mass. 01985. Wholesalers of general nursery stock. *(page 12)*

City Gardener. 105 West 55 Street, New York City 10019.

* Page of first mention in text

Consulting firm for horticultural design. No retail shop. *(page 208)*

James J. Coghlan. Saddle River, N.J. *Out of business in 1971.* *(page 179)*

Conard-Pyle Company. West Grove, Penn. 19390. *(page 74)*

Cooley's Gardens. Silverton, Ore. 97381. *(page 98)*

P. de Jager & Sons. Now de Jager Bulbs, Inc., 188 Asbury Street, S. Hamilton, Mass. 01982. *(page 13)*

Edward H. Scanlon & Assoc. 7621 Lewis Road, Olmsted Falls, Ohio 44138. *(page 111)*

Fa. D. Bruidegom. Baarn, Holland. *Out of business.* *(page 96)*

Ferry-Morse Seed Company. P.O. Box 100, Mountain View, Calif. 94042. *(page 143)*

Gardens of the Blue Ridge. P.O. Box 10, Pineola, N.C. 28662. *(page 119)*

Gardenside Nurseries. Shelburne, Vt. 05482. Now deal in nursery stock and garden center landscaping. *(page 40)*

George, James (*see* James I. George & Son)

Geo. W. Park Seed Company. Box 31, Greenwood, S.C. 29647. *(page 22)*

Gilbert H. Wild & Son. Sarcoxie, Mo. 64862 *(page 121)*

Girard Nurseries. Geneva, Ohio 44041. *(page 182)*

Giridlian, J. N. (*see* Oakhurst Gardens)

Harris, Joseph (*see* Joseph Harris Company)

J. Heemskerk. Now distributed by C. M. van Trigt, Windmill Garden Centers, Abingdon, Md. 21009. *(page 123)*

Roy Hennessey. Scappoose, Ore. 97056. No catalogue since 1962; no listing in local directories. *(page 36)*

Horticulture, Inc. Flowerama no longer produced. *(page 30)*

Cecil Houdyshel. La Verne, Calif. *Out of business.* *(page 21)*

Jackson & Perkins Company. 1 Rose Lane, Medford, Ore. 97501. *(page 9)*

James I. George & Son. Fairport, N.Y. 14450. *(page 92)*

John Scheepers, Inc. 63 Wall Street, New York City 10005. *(page 13)*

Joseph Harris Company. Moreton Farm, Rochester, N.Y. 14626. *(page 3)*

Joseph J. Kern Rose Nursery. Box 33, Mentor, Ohio 44060. *(page 79)*

Kelly Brothers Nurseries, Inc. Dansville, N.Y. 14437. *(page 108)*

Kelsey Nursery Service. Highlands, N.J. *Out of business in 1969.* *(page 107)*

Lamb Nurseries. East 101 Sharp Avenue, Spokane, Wash. 99202. *(page 93)*

Henry Leuthardt. Montauk Highway, East Moriches, N.Y. 11940. *(page 175)*

Lindum Gardens 1309 SW Washington St., Portland, Ore. 97302. No catalogue since 1961; no listing in local directories. *(page 109)*

Little England Daffodil Farm. Gloucester County, Va. 23061. *(page 124)*

Lounsberry Gardens. Oakford, Ill. 62673. *(page 117)*

Mandeville & King Company. Rochester, N.Y. 14626. *(page 145)*

Max Schling Seedsmen. Now Max Schling Inc., florist. 470 Park Avenue, New York City 10022. No longer sells flowering bulbs. *(page 13)*

Mayfair Nurseries. Nichols, N.Y. *Out of business.* *(page 181)*

Merry Gardens. Camden, Me. 04843. *(page 56)*

National Farm Equipment Company. 645 Broadway, New York City 10012. *(page 170)*

Oakhurst Gardens. Box 444, Arcadia, Calif. 91006. No catalogue since 1968; no listing in local directories. *(page 38)*

Oregon Bulb Farms. Gresham, Ore. 97030. *(page 65)*

Park, George (*see* George W. Park Seed Co., Inc.)

Parrella Dahlia Gardens. The Bronx, N.Y. *Out of business in 1968.* *(page 97)*

Pearce Seed Company. Moorestown, N.J. *Out of business in 1969.* *(page 16)*

Peterson & Dering. Scappoose, Ore. *Out of business.* *(page 152)*

Amos Pettingill (*see* White Flower Farm)

Pitzonka's Pansy Farm and Nursery. 3101 Oxford Valley Road, Levittown, Penn. 19057. General nursery stock. *(page 40)*

Putney Nursery. Putney, Vt. 05346. *(page 41)*

Ransom Seed Company. Arcadia, Calif. 91006; San Gabriel Calif. 91006. No listing in local directories. *(page 98)*

Rayner Brothers. Salisbury, Md. 21801. *(page 178)*

Roses of Yesterday and Today (*see* Will Tillotson's Roses)

H. M. Russell. Russell Gardens, Spring, Tex. *Out of business.* *(page 5)*

Harry E. Saier. Dimondale, Mich. Mr. Saier now deceased. Work carried on by J. L. Hudson, Box 1058, Redwood City, Calif. 94064. *(page 25)*

A. P. Saunders. Clinton, N.Y. 13323. Died in 1953. Daughter Silvia Saunders continued business until 1976, promoting hybrids her father developed. Dr. David Reath, Vulcan, Mich. 49812, now carries almost all of Saunders's peony hybrids. *(page 12)*

Scanlon, Edward (*see* Edward H. Scanlon & Assoc.)

Scheepers, John (*see* John Scheepers, Inc.)

S. Scherer & Sons. Northport, N.Y. 11768. *(page 132)*

Schling (*see* Max Schling Seedsmen)

Slocum Water Gardens. 1101 Cypress Gardens Road, Winter Haven, Fla. 33880. *(page 132)*

Strawberry Hill Nursery. Rhinebeck, N.Y. *Out of business in 1961.* *(page 112)*

Swan Island Dahlias. Canby, Ore. 97013. *(page 96)*

Thomasville Nurseries. Thomasville, Ga. 31792. *(page 81)*

Three Springs Fisheries. Lilypons, Md. 21717. *(page 131)*

Tillotson (*see* Will Tillotson's Roses)

Tinari Greenhouses. 2325 Valley Road, Huntingdon, Penn. 19006. *(page 60)*

Tricker, William (*see* William Tricker, Inc.)

Van Ness Water Gardens. 2460 North Euclid Avenue, Upland, Calif. 91786. *(page 131)*

Vaughan's Seed Company. Now a division of Vaughan-Jacklin Corp., wholesale broker. Vaughan Seed Company's northeast office is Chimney Rock Road, Bound Brook, N.J. 08805. Main office—5300 Katrine Avenue, Downers Grove, Ill. 60515. *(page 29)*

Vetterle & Reinelt. Capitola, Calif. *Out of business. (page 60)*

Village Hill Nursery. Williamsburg, Mass. *Out of business in 1965. (page 40)*

W. Atlee Burpee (*see* Burpee)

Wake Robin Farm. Home, Penn. *Out of business in 1968. (page 16)*

Romaine B. Ware. Canby, Ore. *Out of business. (page 13)*

Wayside Gardens. Hodges, S.C. 29653. *(page 10)*

Weston Nurseries. Hopkinton, Mass. 01748. *(page 106)*

White Flower Farm. Litchfield, Conn. 06759. New owner, Eliot Wadsworth II, now author of *White Flower Farm Notes.* William Harris, former owner, now retired but does free-lance writing. *(page 6)*

Wild, Gilbert (*see* Gilbert H. Wild & Son)

Will Tillotson's Roses. Now "Roses of Yesterday and Today," Brown's Valley Road, Watsonville, Calif. 95076. Mr. Tillotson willed business to his assistant, Dorothy C. Stemler, who ran it until her death in 1976. Patricia Stemler Wiley, her daughter, now runs the business. *(page 8)*

William Tricker, Inc. Saddle River, N.J. 07458. *(page 132)*

Melvin E. Wyant. Johnny Cake Ridge, Mentor, Ohio 44060. *(page 78)*

Books in Print

Anderson, Frank J. *The Complete Book of 168 Redouté Roses.* New York: Abbeville Press, 1979. (KSW mentions only the Ariel Press edition, now unavailable.) *(page 81*)*

Bailey, Liberty H. *Hortus Third.* New York: Macmillan, Inc., 1976. (KSW mentions only Bailey's *Hortus Second*, now unavailable.) *(page 167)*

Ballard, Ernesta Drinker. *Garden in Your House.* Rev. and enlarged ed. New York: Harper & Row, Publishers, Inc., 1971. *(page 47)*

———. *The Art of Training Plants.* New York: Barnes & Noble Books, 1974. *(page 180)*

Baskin, Leonard & Esther. *The Poppy and Other Deadly Plants.* New York: Delacorte Press, 1967. *(page 297)*

Bruce, Harold. *Winterthur in Bloom: Winter, Spring, Summer, Autumn.* Wilmington: Winterthur, 1968. *(page 313)*

Campbell, Carlos C., et al. *Great Smoky Mountains Wildflowers.* 4th ed. Knoxville: University of Tennessee Press, 1977. *(page 320)*

Dana, Frances T. *How to Know the Wild Flowers.* New York: Dover Publications, Inc., 1963. *(page 175)*

Dormon, Caroline. *Flowers Native to the Deep South.* Baton Rouge: Claitor's Law Books and Publishing, 1958. *(page 119)*

Dutton, Joan Parry. *The Flower World of Williamsburg.* Rev. ed. Williamsburg: Colonial Williamsburg, 1973. *(page 256)*

Farrer, Reginald. *The English Rock-Garden.* 2 vols. Reprint of 1919 ed. Little Compton, Rhode Island: Theophrastus, 1976. *(page 321)*

Gerard, John. *The Herbal or General History of Plants.* Reprint

* Page of first mention in text

of 1633 ed. New York: Dover Publications, Inc., 1975. *(page 334)*

Hay, Roy & Patrick M. Synge. *The Color Dictionary of Flowers and Plants for Home and Garden.* New York: Crown Publishers, Inc., 1975. *(page 344)*

Hearn, Lafcadio. *Glimpses of Unfamiliar Japan.* 2 vols. Reprint of 1874 ed. St. Clair Shores, Michigan: Scholarly Press, 1974; and Rutland, Vermont: Charles E. Tuttle Co., Inc., 1976. *(page 249)*

House, Homer D. *Wild Flowers.* New York: Macmillan, Inc., 1974. *(page 197)*

Hollingsworth, Buckner. *Gardening on Main Street.* New Brunswick, New Jersey: Rutgers University Press, 1968. *(page 322)*

Jacob, Dorothy. *A Witch's Guide to Gardening.* New York: Taplinger Publishing Co., Inc., 1965. *(page 216)*

Jekyll, Gertrude. *Old English Household Life.* New York: Hippocrene Books, Inc., 1975. (KSW does not mention this title; it is the only book by Jekyll now available.) *(page 114)*

Justice, Wm. S. & C. Ritchie Bell. *Wild Flowers of North Carolina.* Chapel Hill: University of North Carolina Press, 1968. *(page 341)*

Kasperski, Victoria R. *How to Make Cut Flowers Last.* New York: William Morrow & Co., Inc., 1975. *(page 288)*

Kauffman, Richard & John Muir. *Gentle Wilderness: The Sierra Nevada.* San Francisco: Sierra Club Books, 1967. *(page 199)*

Kerr, Jessica. *Shakespeare's Flowers.* New York: Thomas Y. Crowell Co., Inc., 1969. *(page 345)*

Kilvert, Francis. Passages from his *Diary* included in *The Poetry of Earth*, an anthology selected by E. D. H. Johnson. New York: Atheneum Publishers, 1966. *(page 261)*

Kingsbury, John M. *Poisonous Plants of the United States and Canada.* Englewood Cliffs, New Jersey: Prentice-Hall, Inc., 1964. *(page 297)*

Krutch, Joseph Wood. *Herbal.* Boston: David R. Godine, Publisher, 1976. *(page 215)*

Lawrence, Elizabeth. *Gardens in Winter.* Baton Rouge: Claitor's Law Books and Publishing, 1973. *(page 134)*

———. *A Southern Garden.* Chapel Hill: University of North Carolina Press, 1967. *(page 324)*

Leighton, Ann. *Early American Gardens,* '*For Meate or Medecine.*' Boston: Houghton Mifflin Company, 1970. *(page 335)*

Liberman, Alexander. *The Artist in His Studio.* New York: Penguin Books, 1974. *(page 257)*

Menninger, Edwin A. *Fantastic Trees.* Stuart, Florida: Horticultural Books, Inc., 1975. *(page 306)*

Nehrling, Arno & Irene. *Peonies, Outdoors and In.* New York: Dover Publications, Inc., 1975; and Magnolia, Massachusetts: Peter Smith, 1976. *(page 120)*

Nichols, Beverly. *The Art of Flower Arrangement.* New York: The Viking Press, 1967. *(page 326)*

Parkinson, John. *A Garden of Pleasant Flowers: Paradisi in Sole, Paradisus Terrestris.* Reprint of 1629 ed. New York: Dover Publications, Inc., 1976. *(page 334)*

———. *Paradisi in Sole, Paradisus Terrestris.* Reprint of 1597 ed. Norwood, New Jersey: Walter J. Johnson, Inc., 1974.

Peterson, Roger Tory & Margaret McKenny. *A Field Guide to Wildflowers of Northeastern and North-Central North America.* Boston: Houghton Mifflin Company, 1977. *(page 319)*

Porter, Eliot & Henry David Thoreau. *In Wilderness Is the Preservation of the World.* San Francisco: Sierra Club Books, 1962; and New York: Ballantine Books, Inc., 1976. *(page 198)*

Redouté, Pierre-Joseph (*see* Anderson)

Rickett, Harold W. *Wild Flowers of the United States*. 6 Vols. & Index. New York: McGraw-Hill Book Co., 1966–1975. *(page 221)*

Rohde, Eleanour S. *The Scented Garden*. Reprint of 1931 ed. Detroit: Gale Research Company, 1974. *(page 301)*

Sackville-West, Vita. *V. Sackville-West's Garden Book*. New York: Atheneum Publishers, 1979. *(page 384)*

Sargent, Charles Sprague. *Manual of the Trees of North America*. 2 vols., 2nd ed. New York: Dover Publications, 1977; and Magnolia, Massachusetts: Peter Smith, 1962. *(page 222)*

———. *Silva of North America*. Magnolia, Massachusetts: Peter Smith, 1979. *(page 221)*

Smith, A. W. & W. T. Williams. *A Gardener's Dictionary of Plant Names: A Handbook on the Origin and Meaning of Some Plant Names*. New York: St. Martin's Press, Inc., 1972. (Original edition, *A Gardener's Book of Plant Names*, now unavailable.) *(page 209)*

Taylor, Norman, ed. *Encyclopedia of Gardening*. Rev. ed. Boston: Houghton Mifflin Company, 1961. *(page 162)*

Truex, Philip. *The City Gardener*. New York: Alfred A. Knopf, Inc., 1963. *(page 208)*

Whittle, Tyler. *Some Ancient Gentlemen*. New York: Taplinger Publishing Co., Inc., 1966. *(page 307)*

Wilder, Louise Beebe. *The Fragrant Garden*. Reprint of 1936 ed. New York: Dover Publications, Inc., 1974. (KSW mentions *The Fragrant Path*, not *The Fragrant Garden*.) *(page 70)*

Wyman, Donald. *Shrubs and Vines for American Gardens*. Rev. and enlarged ed. New York: Macmillan Publishing Co., Inc., 1969. *(page 113)*

———. *The Saturday Morning Gardener*. Rev. ed. New York: Macmillan Publishing Co., Inc., 1974. *(page 209)*

OTHER NEW YORK REVIEW CLASSICS

For a complete list of titles, visit www.nyrb.com or write to: Catalog Requests, NYRB, 435 Hudson Street, New York, NY 10014

J.R. ACKERLEY Hindoo Holiday*
J.R. ACKERLEY My Dog Tulip*
J.R. ACKERLEY My Father and Myself*
J.R. ACKERLEY We Think the World of You*
HENRY ADAMS The Jeffersonian Transformation
RENATA ADLER Pitch Dark*
RENATA ADLER Speedboat*
CÉLESTE ALBARET Monsieur Proust
DANTE ALIGHIERI The Inferno
DANTE ALIGHIERI The New Life
KINGSLEY AMIS The Alteration*
KINGSLEY AMIS Ending Up*
KINGSLEY AMIS Girl, 20*
KINGSLEY AMIS The Green Man*
KINGSLEY AMIS Lucky Jim*
KINGSLEY AMIS The Old Devils*
KINGSLEY AMIS One Fat Englishman*
KINGSLEY AMIS Take a Girl Like You*
WILLIAM ATTAWAY Blood on the Forge
W.H. AUDEN (EDITOR) The Living Thoughts of Kierkegaard
W.H. AUDEN W.H. Auden's Book of Light Verse
ERICH AUERBACH Dante: Poet of the Secular World
DOROTHY BAKER Cassandra at the Wedding*
DOROTHY BAKER Young Man with a Horn*
J.A. BAKER The Peregrine
S. JOSEPHINE BAKER Fighting for Life*
HONORÉ DE BALZAC The Human Comedy: Selected Stories*
HONORÉ DE BALZAC The Unknown Masterpiece *and* Gambara*
SYBILLE BEDFORD A Legacy*
MAX BEERBOHM Seven Men
STEPHEN BENATAR Wish Her Safe at Home*
FRANS G. BENGTSSON The Long Ships*
ALEXANDER BERKMAN Prison Memoirs of an Anarchist
GEORGES BERNANOS Mouchette
ADOLFO BIOY CASARES Asleep in the Sun
ADOLFO BIOY CASARES The Invention of Morel
CAROLINE BLACKWOOD Great Granny Webster*
NICOLAS BOUVIER The Way of the World
MALCOLM BRALY On the Yard*
MILLEN BRAND The Outward Room*
SIR THOMAS BROWNE Religio Medici and Urne-Buriall*
JOHN HORNE BURNS The Gallery
ROBERT BURTON The Anatomy of Melancholy
CAMARA LAYE The Radiance of the King
GIROLAMO CARDANO The Book of My Life
DON CARPENTER Hard Rain Falling*
J.L. CARR A Month in the Country*

* *Also available as an electronic book.*

BLAISE CENDRARS Moravagine
EILEEN CHANG Love in a Fallen City
JOAN CHASE During the Reign of the Queen of Persia*
UPAMANYU CHATTERJEE English, August: An Indian Story
NIRAD C. CHAUDHURI The Autobiography of an Unknown Indian
ANTON CHEKHOV Peasants and Other Stories
GABRIEL CHEVALLIER Fear: A Novel of World War I*
RICHARD COBB Paris and Elsewhere
COLETTE The Pure and the Impure
JOHN COLLIER Fancies and Goodnights
CARLO COLLODI The Adventures of Pinocchio*
IVY COMPTON-BURNETT A House and Its Head
IVY COMPTON-BURNETT Manservant and Maidservant
BARBARA COMYNS The Vet's Daughter
EVAN S. CONNELL The Diary of a Rapist
ALBERT COSSERY The Jokers*
ALBERT COSSERY Proud Beggars*
HAROLD CRUSE The Crisis of the Negro Intellectual
ASTOLPHE DE CUSTINE Letters from Russia*
LORENZO DA PONTE Memoirs
ELIZABETH DAVID A Book of Mediterranean Food
ELIZABETH DAVID Summer Cooking
L.J. DAVIS A Meaningful Life*
VIVANT DENON No Tomorrow/Point de lendemain
MARIA DERMOÛT The Ten Thousand Things
DER NISTER The Family Mashber
TIBOR DÉRY Niki: The Story of a Dog
ARTHUR CONAN DOYLE The Exploits and Adventures of Brigadier Gerard
CHARLES DUFF A Handbook on Hanging
BRUCE DUFFY The World As I Found It*
DAPHNE DU MAURIER Don't Look Now: Stories
ELAINE DUNDY The Dud Avocado*
ELAINE DUNDY The Old Man and Me*
G.B. EDWARDS The Book of Ebenezer Le Page
JOHN EHLE The Land Breakers*
EURIPIDES Grief Lessons: Four Plays; translated by Anne Carson
J.G. FARRELL Troubles*
J.G. FARRELL The Siege of Krishnapur*
J.G. FARRELL The Singapore Grip*
ELIZA FAY Original Letters from India
KENNETH FEARING The Big Clock
KENNETH FEARING Clark Gifford's Body
FÉLIX FÉNÉON Novels in Three Lines*
M.I. FINLEY The World of Odysseus
THOMAS FLANAGAN The Year of the French*
SANFORD FRIEDMAN Conversations with Beethoven*
SANFORD FRIEDMAN Totempole*
MASANOBU FUKUOKA The One-Straw Revolution*
MARC FUMAROLI When the World Spoke French
CARLO EMILIO GADDA That Awful Mess on the Via Merulana
BENITO PÉREZ GÁLDOS Tristana*
MAVIS GALLANT The Cost of Living: Early and Uncollected Stories*

MAVIS GALLANT Paris Stories*
MAVIS GALLANT Varieties of Exile*
GABRIEL GARCÍA MÁRQUEZ Clandestine in Chile: The Adventures of Miguel Littín
ALAN GARNER Red Shift*
WILLIAM H. GASS In the Heart of the Heart of the Country: And Other Stories*
WILLIAM H. GASS On Being Blue: A Philosophical Inquiry*
THÉOPHILE GAUTIER My Fantoms
JEAN GENET Prisoner of Love
ÉLISABETH GILLE The Mirador: Dreamed Memories of Irène Némirovsky by Her Daughter*
JOHN GLASSCO Memoirs of Montparnasse*
P.V. GLOB The Bog People: Iron-Age Man Preserved
NIKOLAI GOGOL Dead Souls*
EDMOND AND JULES DE GONCOURT Pages from the Goncourt Journals
PAUL GOODMAN Growing Up Absurd: Problems of Youth in the Organized Society*
EDWARD GOREY (EDITOR) The Haunted Looking Glass
JEREMIAS GOTTHELF The Black Spider*
A.C. GRAHAM Poems of the Late T'ang
WILLIAM LINDSAY GRESHAM Nightmare Alley*
EMMETT GROGAN Ringolevio: A Life Played for Keeps
VASILY GROSSMAN An Armenian Sketchbook*
VASILY GROSSMAN Everything Flows*
VASILY GROSSMAN Life and Fate*
VASILY GROSSMAN The Road*
OAKLEY HALL Warlock
PATRICK HAMILTON The Slaves of Solitude
PATRICK HAMILTON Twenty Thousand Streets Under the Sky
PETER HANDKE Short Letter, Long Farewell
PETER HANDKE Slow Homecoming
ELIZABETH HARDWICK The New York Stories of Elizabeth Hardwick*
ELIZABETH HARDWICK Seduction and Betrayal*
ELIZABETH HARDWICK Sleepless Nights*
L.P. HARTLEY Eustace and Hilda: A Trilogy*
L.P. HARTLEY The Go-Between*
NATHANIEL HAWTHORNE Twenty Days with Julian & Little Bunny by Papa
ALFRED HAYES In Love*
ALFRED HAYES My Face for the World to See*
PAUL HAZARD The Crisis of the European Mind: 1680–1715*
GILBERT HIGHET Poets in a Landscape
RUSSELL HOBAN Turtle Diary*
HUGO VON HOFMANNSTHAL The Lord Chandos Letter*
JAMES HOGG The Private Memoirs and Confessions of a Justified Sinner
RICHARD HOLMES Shelley: The Pursuit*
ALISTAIR HORNE A Savage War of Peace: Algeria 1954–1962*
GEOFFREY HOUSEHOLD Rogue Male*
WILLIAM DEAN HOWELLS Indian Summer
BOHUMIL HRABAL Dancing Lessons for the Advanced in Age*
DOROTHY B. HUGHES The Expendable Man*
RICHARD HUGHES A High Wind in Jamaica*
RICHARD HUGHES In Hazard*
INTIZAR HUSAIN Basti*
MAUDE HUTCHINS Victorine
YASUSHI INOUE Tun-huang*

HENRY JAMES The Ivory Tower
HENRY JAMES The New York Stories of Henry James*
HENRY JAMES The Other House
HENRY JAMES The Outcry
TOVE JANSSON Fair Play *
TOVE JANSSON The Summer Book*
TOVE JANSSON The True Deceiver*
TOVE JANSSON The Woman Who Borrowed Memories: Selected Stories*
RANDALL JARRELL (EDITOR) Randall Jarrell's Book of Stories
DAVID JONES In Parenthesis
JOSEPH JOUBERT The Notebooks of Joseph Joubert; translated by Paul Auster
KABIR Songs of Kabir; translated by Arvind Krishna Mehrotra*
FRIGYES KARINTHY A Journey Round My Skull
ERICH KÄSTNER Going to the Dogs: The Story of a Moralist*
HELEN KELLER The World I Live In
YASHAR KEMAL Memed, My Hawk
MURRAY KEMPTON Part of Our Time: Some Ruins and Monuments of the Thirties*
RAYMOND KENNEDY Ride a Cockhorse*
DAVID KIDD Peking Story*
ARUN KOLATKAR Jejuri
DEZSŐ KOSZTOLÁNYI Skylark*
TÉTÉ-MICHEL KPOMASSIE An African in Greenland
GYULA KRÚDY The Adventures of Sindbad*
GYULA KRÚDY Sunflower*
SIGIZMUND KRZHIZHANOVSKY Autobiography of a Corpse*
SIGIZMUND KRZHIZHANOVSKY The Letter Killers Club*
SIGIZMUND KRZHIZHANOVSKY Memories of the Future
GIUSEPPE TOMASI DI LAMPEDUSA The Professor and the Siren
MARGARET LEECH Reveille in Washington: 1860–1865*
PATRICK LEIGH FERMOR Between the Woods and the Water*
PATRICK LEIGH FERMOR The Broken Road*
PATRICK LEIGH FERMOR Mani: Travels in the Southern Peloponnese*
PATRICK LEIGH FERMOR Roumeli: Travels in Northern Greece*
PATRICK LEIGH FERMOR A Time of Gifts*
PATRICK LEIGH FERMOR A Time to Keep Silence*
PATRICK LEIGH FERMOR The Traveller's Tree*
D.B. WYNDHAM LEWIS AND CHARLES LEE (EDITORS) The Stuffed Owl
SIMON LEYS The Hall of Uselessness: Collected Essays
GEORG CHRISTOPH LICHTENBERG The Waste Books
JAKOV LIND Soul of Wood and Other Stories
DWIGHT MACDONALD Masscult and Midcult: Essays Against the American Grain*
NORMAN MAILER Miami and the Siege of Chicago*
CURZIO MALAPARTE Kaputt
CURZIO MALAPARTE The Skin
JANET MALCOLM In the Freud Archives
JEAN-PATRICK MANCHETTE Fatale*
JEAN-PATRICK MANCHETTE The Mad and the Bad*
OSIP MANDELSTAM The Selected Poems of Osip Mandelstam
OLIVIA MANNING Fortunes of War: The Balkan Trilogy*
OLIVIA MANNING Fortunes of War: The Levant Trilogy*
OLIVIA MANNING School for Love*
JAMES VANCE MARSHALL Walkabout*

GUY DE MAUPASSANT Afloat
GUY DE MAUPASSANT Alien Hearts*
JAMES McCOURT Mawrdew Czgowchwz*
WILLIAM McPHERSON Testing the Current*
DAVID MENDEL Proper Doctoring: A Book for Patients and Their Doctors*
HENRI MICHAUX Miserable Miracle
JESSICA MITFORD Hons and Rebels
JESSICA MITFORD Poison Penmanship*
NANCY MITFORD Frederick the Great*
NANCY MITFORD Madame de Pompadour*
NANCY MITFORD Voltaire in Love*
MICHEL DE MONTAIGNE Shakespeare's Montaigne; translated by John Florio*
HENRY DE MONTHERLANT Chaos and Night
BRIAN MOORE The Lonely Passion of Judith Hearne*
BRIAN MOORE The Mangan Inheritance*
ALBERTO MORAVIA Agostino*
ALBERTO MORAVIA Boredom*
ALBERTO MORAVIA Contempt*
JAN MORRIS Conundrum*
JAN MORRIS Hav*
PENELOPE MORTIMER The Pumpkin Eater*
ÁLVARO MUTIS The Adventures and Misadventures of Maqroll
NESCIO Amsterdam Stories*
DARCY O'BRIEN A Way of Life, Like Any Other
SILVINA OCAMPO Thus Were Their Faces*
YURI OLESHA Envy*
IONA AND PETER OPIE The Lore and Language of Schoolchildren
IRIS OWENS After Claude*
RUSSELL PAGE The Education of a Gardener
ALEXANDROS PAPADIAMANTIS The Murderess
BORIS PASTERNAK, MARINA TSVETAYEVA, AND RAINER MARIA RILKE Letters, Summer 1926
CESARE PAVESE The Moon and the Bonfires
CESARE PAVESE The Selected Works of Cesare Pavese
LUIGI PIRANDELLO The Late Mattia Pascal
JOSEP PLA The Gray Notebook
ANDREY PLATONOV The Foundation Pit
ANDREY PLATONOV Happy Moscow
ANDREY PLATONOV Soul and Other Stories
J.F. POWERS Morte d'Urban*
J.F. POWERS The Stories of J.F. Powers*
J.F. POWERS Wheat That Springeth Green*
CHRISTOPHER PRIEST Inverted World*
BOLESŁAW PRUS The Doll*
ALEXANDER PUSHKIN The Captain's Daughter*
QIU MIAOJIN Last Words from Montmartre*
RAYMOND QUENEAU We Always Treat Women Too Well
RAYMOND QUENEAU Witch Grass
FRIEDRICH RECK Diary of a Man in Despair*
JULES RENARD Nature Stories*
JEAN RENOIR Renoir, My Father
GREGOR VON REZZORI An Ermine in Czernopol*
GREGOR VON REZZORI Memoirs of an Anti-Semite*

GREGOR VON REZZORI The Snows of Yesteryear: Portraits for an Autobiography*
TIM ROBINSON Stones of Aran: Labyrinth
TIM ROBINSON Stones of Aran: Pilgrimage
MILTON ROKEACH The Three Christs of Ypsilanti*
FR. ROLFE Hadrian the Seventh
GILLIAN ROSE Love's Work
WILLIAM ROUGHEAD Classic Crimes
CONSTANCE ROURKE American Humor: A Study of the National Character
SAKI The Unrest-Cure and Other Stories; illustrated by Edward Gorey
TAYEB SALIH Season of Migration to the North
JEAN-PAUL SARTRE We Have Only This Life to Live: Selected Essays. 1939–1975
GERSHOM SCHOLEM Walter Benjamin: The Story of a Friendship*
DANIEL PAUL SCHREBER Memoirs of My Nervous Illness
JAMES SCHUYLER Alfred and Guinevere
JAMES SCHUYLER What's for Dinner?*
SIMONE SCHWARZ-BART The Bridge of Beyond*
LEONARDO SCIASCIA The Day of the Owl
LEONARDO SCIASCIA Equal Danger
LEONARDO SCIASCIA The Moro Affair
LEONARDO SCIASCIA To Each His Own
LEONARDO SCIASCIA The Wine-Dark Sea
VICTOR SEGALEN René Leys*
ANNA SEGHERS Transit*
PHILIPE-PAUL DE SÉGUR Defeat: Napoleon's Russian Campaign
GILBERT SELDES The Stammering Century*
VICTOR SERGE The Case of Comrade Tulayev*
VICTOR SERGE Conquered City*
VICTOR SERGE Memoirs of a Revolutionary
VICTOR SERGE Midnight in the Century*
VICTOR SERGE Unforgiving Years
SHCHEDRIN The Golovlyov Family
ROBERT SHECKLEY The Store of the Worlds: The Stories of Robert Sheckley*
GEORGES SIMENON Act of Passion*
GEORGES SIMENON Dirty Snow*
GEORGES SIMENON Monsieur Monde Vanishes*
GEORGES SIMENON Pedigree*
GEORGES SIMENON Three Bedrooms in Manhattan*
GEORGES SIMENON Tropic Moon*
GEORGES SIMENON The Widow*
CHARLES SIMIC Dime-Store Alchemy: The Art of Joseph Cornell
TESS SLESINGER The Unpossessed: A Novel of the Thirties*
VLADIMIR SOROKIN Ice Trilogy*
VLADIMIR SOROKIN The Queue
NATSUME SŌSEKI The Gate*
DAVID STACTON The Judges of the Secret Court*
JEAN STAFFORD The Mountain Lion
GEORGE R. STEWART Names on the Land
STENDHAL The Life of Henry Brulard
ADALBERT STIFTER Rock Crystal
THEODOR STORM The Rider on the White Horse
JEAN STROUSE Alice James: A Biography*
HOWARD STURGIS Belchamber

ITALO SVEVO As a Man Grows Older
A.J.A. SYMONS The Quest for Corvo
MAGDA SZABÓ The Door*
ANTAL SZERB Journey by Moonlight*
ELIZABETH TAYLOR Angel*
ELIZABETH TAYLOR A Game of Hide and Seek*
ELIZABETH TAYLOR You'll Enjoy It When You Get There: The Stories of Elizabeth Taylor*
HENRY DAVID THOREAU The Journal: 1837–1861*
ALEKSANDAR TIŠMA The Use of Man*
TATYANA TOLSTAYA The Slynx
TATYANA TOLSTAYA White Walls: Collected Stories
EDWARD JOHN TRELAWNY Records of Shelley, Byron, and the Author
LIONEL TRILLING The Liberal Imagination*
THOMAS TRYON The Other*
IVAN TURGENEV Virgin Soil
JULES VALLÈS The Child
RAMÓN DEL VALLE-INCLÁN Tyrant Banderas*
MARK VAN DOREN Shakespeare
ELIZABETH VON ARNIM The Enchanted April*
EDWARD LEWIS WALLANT The Tenants of Moonbloom
ROBERT WALSER Berlin Stories*
ROBERT WALSER Jakob von Gunten
ROBERT WALSER A Schoolboy's Diary and Other Stories*
REX WARNER Men and Gods
SYLVIA TOWNSEND WARNER Lolly Willowes*
SYLVIA TOWNSEND WARNER Mr. Fortune*
SYLVIA TOWNSEND WARNER Summer Will Show*
ALEKSANDER WAT My Century*
C.V. WEDGWOOD The Thirty Years War
SIMONE WEIL On the Abolition of All Political Parties*
SIMONE WEIL AND RACHEL BESPALOFF War and the Iliad
GLENWAY WESCOTT The Pilgrim Hawk*
REBECCA WEST The Fountain Overflows
EDITH WHARTON The New York Stories of Edith Wharton*
KATHARINE S. WHITE Onward and Upward in the Garden
PATRICK WHITE Riders in the Chariot
T.H. WHITE The Goshawk*
JOHN WILLIAMS Augustus*
JOHN WILLIAMS Butcher's Crossing*
JOHN WILLIAMS Stoner*
ANGUS WILSON Anglo-Saxon Attitudes
EDMUND WILSON Memoirs of Hecate County
RUDOLF AND MARGARET WITTKOWER Born Under Saturn
GEOFFREY WOLFF Black Sun*
FRANCIS WYNDHAM The Complete Fiction
JOHN WYNDHAM The Chrysalids
BÉLA ZOMBORY-MOLDOVÁN The Burning of the World: A Memoir of 1914*
STEFAN ZWEIG Beware of Pity*
STEFAN ZWEIG Chess Story*
STEFAN ZWEIG Confusion *
STEFAN ZWEIG Journey Into the Past*
STEFAN ZWEIG The Post-Office Girl*